To Ken

With Be

Steve Blaha

A Finite Unified Quantum Field Theory of the Elementary Particle Standard Model and Quantum Gravity

Based on New Quantum Dimensions™ & a New Paradigm in the Calculus of Variations

Some Other Books By Stephen Blaha

Cosmos and Consciousness Second Edition (ISBN: 0972079548, Pingree-Hill Publishing, Auburn, NH, 2002)

The Life Cycle of Civilizations (ISBN: 0972079580, Pingree-Hill Publishing, Auburn, NH, 2002)

A Scientific Inquiry into the Nature of God, the Spiritual, and Near Death Experiences (ISBN: 0972079521, Pingree-Hill Publishing, Auburn, NH, 2002)

A Java™ Programming Introductory and Intermediate Course Second Edition (ISBN: 097207953X, Pingree-Hill Publishing, Auburn, NH, 2002)

C++ for Professional Programming With PC and Unix Applications (ISBN: 1850328013, International Thomson Publishing, Boston, 1995)

The Reluctant Prophets: Has Science Found God? (ISBN: 0759663041, 1stBooks Library, Bloomington, IN, 2001)

The Rhythms of History: A Universal Theory of Civilizations (ISBN: 0972079572, Pingree-Hill Publishing, Auburn, NH, 2002)

Available on bn.com and amazon.com as well as bookstores (please tell clerk to order these Print On Demand books from Ingram distributors).

Cover Credits

A Finite Unified Quantum Field Theory of the Elementary Particle Standard Model and Quantum Gravity

Based on New Quantum Dimensions™ & a New Paradigm in the Calculus of Variations

By

Stephen Blaha, Ph.D.

Pingree-Hill Publishing

Pingree-Hill Publishing
P. O. Box 368
Auburn, NH 03032

ISBN: 0-9720795-9-9

This book is printed on acid free paper.

rev. 00/00/01

To Margaret

CONTENTS

LIST OF FIGURES

PREFACE

After a half century of major progress a successful theory that unifies three of the four known interactions emerged in the 1970's. This renormalizable theory, called the Standard Model, has been remarkably successful in accounting for experimental data on the electromagnetic, weak and strong interactions. While there are issues suggesting that it is an interim theory such as the large number of parameters and particles, and the partial nature of the unification of the electroweak and strong interactions, nevertheless it describes experimental data correctly including data at the highest accessible accelerator energies. Thus it must be taken seriously as the current description of these interactions.

The gravitational interaction is a different story. Attempts at a quantum field theory of gravity have failed because of the appearance of infinities in calculations that cannot be removed by the known techniques of renormalization that have worked in the case of the Standard Model. And the combination of quantum gravity with the Standard Model does not remove the renormalization issues in the gravitational sector. Thus the divergences in Quantum Gravity, as it is usually formulated, have stood in the way of a unified quantum field theory of the known interactions.

The creation of a unified field theory of all known interactions would enable calculations to be performed that might help elucidate the beginning moments of the universe in the first instants of time after the Big Bang. It might also introduce additional features that might be experimentally accessible either now or in the near future. Lastly, it would be a stepping stone to the next generation of (perhaps more intellectually satisfying) fundamental theories of physics.

Superstring theory does appear to achieve a unification (or potential unification) of the known forces in a renormalizable theory. Unfortunately, after roughly thirty years of development the additional particles predicted by Superstring theories have not been found. In addition a direct connection to the Standard Model is lacking. Ideally the higher dimension, *correct*, Superstring theory (if it exists) should approximate the Standard Model at currently accessible experimental energies. This approximation is lacking. Thus we achieve unification in a renormalizable theory at the price of many new unobserved particles and a missing connection to successful current theories.

This book describes a unified *finite* quantum field theory of the four known interactions with a manifest connection to the Standard Model and classical gravitational field theory. Applying Ockham's Razor we develop a theory that is minimally different from current theories at "low energy" but without the infinities normally appearing in perturbative quantum field theory. The unified theory includes a finite quantum gravity. We require that the theory be a close approximation to the Standard Model in its "low energy" limit – a form of *Correspondence Principle*. Along the way we will develop an apparently new problem area, or paradigm, in the Calculus of Variations – the determination of an extremum on a manifold that is itself determined by an extremum.

1. Some Issues Facing Elementary Particle Physics Today

Form of Physically Acceptable Theories

The currently known theories of fundamental interactions and elementary particles fall into two broad categories: conventional quantum field theories and string-based theories that are united with supersymmetry to form superstring theories. Most physicists believe that only theories of these types can meet the reasonable physical requirements of Lorentz invariance and unitarity (positive probabilities summing to unity) while accounting for spin and internal quantum numbers.

It appears that a third form of acceptable theory of elementary particles is possible. This type of theory is a quantum field theory with a string-like (more properly, brane-like) substructure. We will investigate a unified quantum field theory of this type. We will treat it seriously as a modified unified theory of the Standard Model and Quantum Gravity. This theory appears to have the following features:

1. Consistency with the current Standard Model in all respects. The Standard Model describes the large distance, "low energy" behavior of the theory.

2. Consistency with the classical Theory of Gravity. Classical gravity is the large distance, "low energy" limit of the theory.

3. General relativistic covariance, Lorentz invariance and unitarity as required in physically acceptable theories.

4. The unified theory is finite – it contains no infinities.

5. All interactions are modified by a short distance, high energy substructure that begins at some energy presumably much above currently accessible energies. The energy scale could be set by the Planck scale (10^{19} GeV) or a much lower energy such as 10^5 GeV (or lower).

6. It allows modifications of the Standard Model such as further unification through broken higher symmetries without losing features 1 – 5 above.

7. It predicts a number of new phenomena at extremely short distances such as ultra-relativistic dilepton resonances consisting of two leptons of the same charge. An example of this

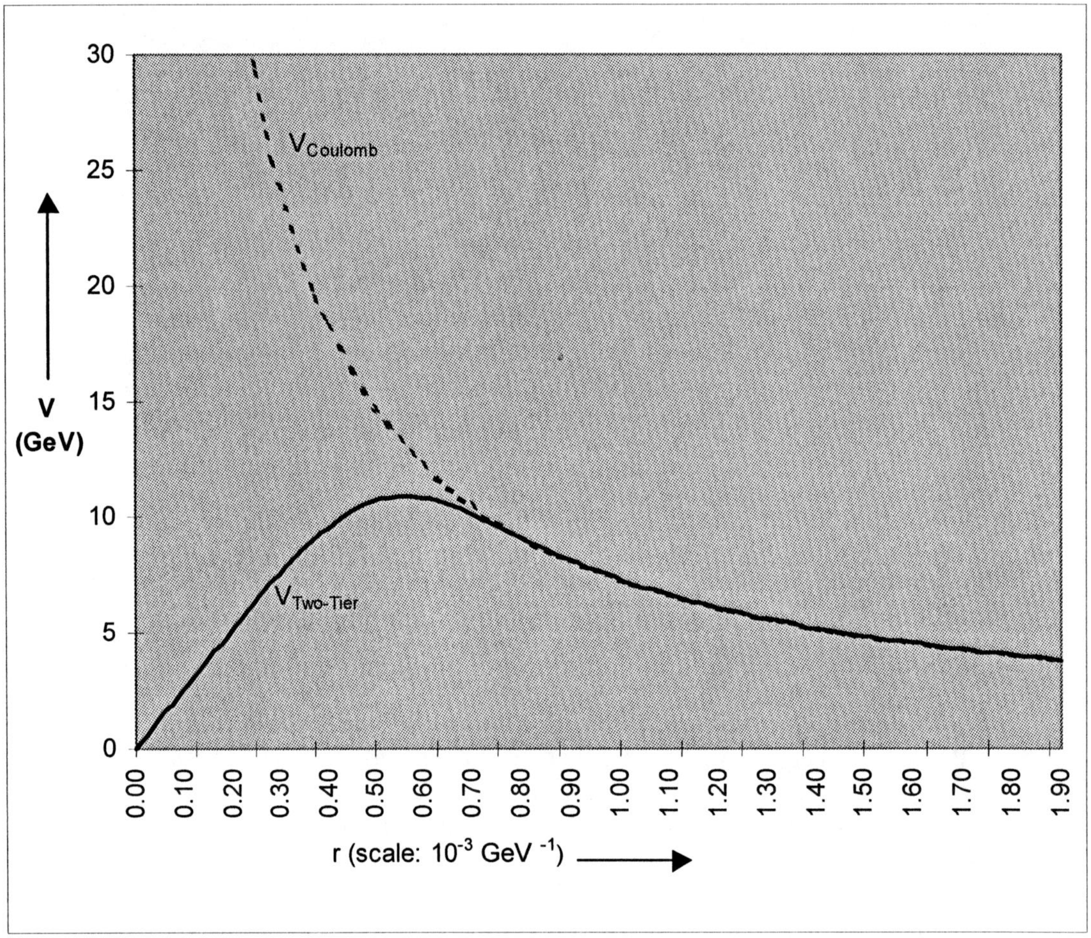

Figure 1.1. Modified electromagnetic Coulomb force between two electrons in the unified theory assuming M_c = 1000 GeV/c^2. The potential in the new theory becomes linear near the origin r = 0 opening up the possibility of an unstable, ultra-relativistic, dilepton bound state. The Compton wavelength of an electron is 1.96×10^3 GeV^{-1} compared to the maximum in V_{new} at r = $.56 \times 10^{-3}$ GeV^{-1} or roughly a million times larger.

type of bound state is a bound resonance consisting of two electrons. These dilepton states could be created at ultrahigh energies by penetrating the repulsive Coulomb barrier.

Inside the barrier near r = 0 the modified Coulomb potential is a linear potential. A dilepton bound state is highly unstable with a large decay rate due to quantum tunneling through the Coulomb barrier.

The curve in Fig. 1.1 is generated from the modified Coulomb force of our new theory for electrons:

$$V_{new} = \alpha\Phi(M_c^2\pi r^2)/r \tag{1.1}$$

where α is the fine structure constant, $\Phi(z)$ is the error function, M_c is the mass setting the scale of the new short distance behavior and r is the radial distance. At small distances ($\pi r^2 \ll M_c^{-2}$)

$$V_{new} \rightarrow \alpha 2\surd\pi\ M_c^2\, r \tag{1.2}$$

and at large distances ($\pi r^2 \gg M_c^{-2}$) it becomes identical to the Coulomb potential:

$$V_{new} \rightarrow V_{Coul} = \alpha/r \tag{1.3}$$

8. The short distance modifications of all four interactions open up the possibility of a more fundamental, smaller set of particles, of which the currently observed particles are bound states. Thus quarks and leptons might be bound states of more fundamental particles.

In the remaining sections of this chapter we consider aspects of the Standard Model and Gravitation, and Superstring theories, that are relevant for the discussion of the new unified theory.

Standard Model

The Standard Model unites the Electroweak interactions and the Strong interactions in a successful theory that appears to be consistent with the known experimental data at all accessible energies. The theory is unitary with positive probabilities adding to unity, Lorentz invariant and renormalizable. There are several variants and proposed extensions of the Standard Model. But the major aspects of the theory are common to all variants.

The most important technical issue that faced the development of the Standard Model was the question of the renormalizability of the unification of the electromagnetic interactions and the weak interactions in the Electroweak Theory sector. The issue was resolved by the proof of renormalizability by ‘t Hooft in 1971.

At first glance Electroweak Theory, and its electromagnetic sector, Quantum Electrodynamics (QED), appear to have divergences when perturbative calculations are made of transition probabilities, or of physical quantities such as the anomalous magnetic moment of the electron or muon. However these divergences can be isolated into a finite number of infinite quantities which in turn can be absorbed into redefinitions of fundamental parameters of the theory (such as the "bare" electric charge or the "bare" mass of the electron.)

For example in QED the the "bare" charge e_0 is renormalized by infinite factors to give the "physical" charge e:

$$e = Z_1^{-1} Z_2 \sqrt{Z_3}\, e_0 \tag{1.4}$$

where Z_1, Z_2, and Z_3 are divergent renormalization constants. Numerous authors over the last fifty years or so have commented on the "unnaturalness" of renormalization which, after all, amounts to multiplying infinity by zero (e_0) to obtain a finite observable number, the electric charge e in this case.

The quantum field theory formalism that we shall develop resolves these issues by being finite – no infinities appear because of a short distance modification that appears in all particle propagators. The result is a logically satisfactory theory that avoids divergences independent of the details of the interactions and symmetries, and, in fact, allows a wider range of interactions such as interactions with derivative couplings, which were previously totally unrenormalizable.

Quantization of Gravity

Classical gravity began with Newton's theory of gravity, which still remains an acceptable theory for most phenomena involving small masses and velocities. In the early twentieth century A. Einstein developed a new theory of gravitation based on geometrical concepts. This theory, the General Theory of Relativity, has been tested and found to be correct as far as we can determine at the classical level.

There have been many attempts to develop a quantum version of the General Theory of Relativity. While part of the overall framework of such a quantum theory is known, all attempts to develop it within the framework of conventional quantum field theory have failed due to the infinite number of divergences that appear when calculations are attempted in perturbation theory.[1] A theory with an infinite number of divergences cannot be handled in the fashion of the divergences of QED, Yang-Mills theories, or the Electroweak Theory; and thus does not have predictive power.

The gravitational sector of the unified theory that we will develop does not have these divergences and is in fact finite.

Correspondence Principle

The theory of Quantum Mechanics was developed using a principle called the *Correspondence Principle.* Simply put, this principle states that quantum mechanical systems must approach the behavior of classical systems as the features of the quantum mechanical

---

[1] See for example B. S. DeWitt, Phys. Rev. **162**, 1239 (1967), and R. P. Feynman, Acta Physica Polonica **24**, 697 (1963).

system become much larger than the quantum of action – Planck's constant h. Classical mechanics is thus the large scale limit of quantum mechanics.

Similarly, the Theory of Special Relativity also has a sort of correspondence principle – the predictions of the theory of Special Relativity for a system must approach the predictions of classical mechanics in the limit that all the velocities of the system become much smaller than the speed of light. Thus classical mechanics is a limiting case of Special Relativity.

Just as these twentieth century theories have a type of correspondence principle in which they must approximate an earlier theory in a certain limit, any new theory of elementary particles must approximate the Standard Model in the currently known range of energies. After all the Standard Model works! Therefore it must have an element of physical truth and a more fundamental theory must embody that truth in the explored range of experimental energies.

Furthermore the extreme accuracy of QED calculations[2] must be matched by any theory purporting to account for elementary particle interactions. At the moment, and apparently for the foreseeable future, Superstring theories have difficulties when attempts are made to find the Standard Model, or something like it, as an approximation at current energies. Capturing the accuracy of QED predictions would seem to be in the distant future, if at all, in Superstring theories.

String and Superstring Theories

W. Pauli once said, "Man should not join together that which God has put asunder." In apparent contradiction (perhaps) to this dictum Superstring theory attempts to unite fermions (half-integer spin particles) and bosons (integer-spin particles) within a larger symmetry so that they can, in a sense, be rotated into one another. While support for this symmetry is currently lacking in elementary particle physics, there may be evidence for supersymmetry in nuclear physics.

However the price for supersymmetry in particle physics appears to be quite high: a large number of dimensions and a large number of particles. No evidence exists at the time of this writing for more than four space-time dimensions or for the supersymmetric partners of the known elementary particles.

Thus, at best, supersymmetry is for the future when experimental energies reach energies where supersymmetric features are unequivocally seen (if in fact they are.)

The theory of strings – two-dimensional substructures that constitute elementary particles – is grounded in phenomenology – more particularly, in the Veneziano-Suzuki formula for scattering amplitudes. Y. Nambu and T. Goto developed a vibrating string theory that accounted for the form of the Veneziano-Suzuki formula. So there appears to be some justification for a string model of elementary particles. In this model the elementary particles, which appear as structure-less and point-like at lower energies, can be seen to have a string-like structure at very high energies.

Thus a sub-structure for elementary particles may have some experimental justification.

[2] T. Kinoshita, "The Fine Structure Constant", Cornell Univ. Preprint CLNS 96/1406 (1996).

Logic Does Not Bring Progress in Physics

An often-expressed hope in the last two decades of the twentieth century was that some unique Superstring theory would emerge, and that this theory would prove to be the only logically reasonable theory. After thirty years of effort by a large group of extremely talented physicists (perhaps larger than any group of physics theorists devoted to one topic since the Manhattan project) this hope is yet to be realized. Historically, the movement from level to deeper level to yet deeper level in Physics have not been the result of logical inquiry but rather have been the result of new experiments confronting existing theory wherein the existing theory is found to be wanting.[3] A rational physicist in the eighteenth century would not have conjectured the theory of special relativity as a logical possibility: "Why should mechanics change when the speed of an object becomes large?" A rational physicist in the nineteenth century would not have conjectured the theory of quantum mechanics as a logical possibility: "Why should it be impossible to measure the momentum and position of a particle with arbitrary precision?" These questions did not, and could not have arisen, in a rational, "down to earth" scientist. Furthermore, despite knowing that these theories are correct we do not know WHY they are implemented in Nature. We only know that they are.

With this historical perspective in mind it appears that we cannot hope to arrive at the form of the next deeper level of Physics through reason alone. Nature will most likely surprise us again (and again). Thus experiment is our teacher as we learn more of the nature of matter and space-time.

Goal: Unified Theory Without Renormalization Issues

There are a number of issues confronting elementary particle physics today. Issues such as CP violation, the nature of dark matter and of dark energy, the form of the symmetries of Nature, and the origin of the "numerous" particles and constants in the Standard Model. Practically speaking, perhaps the most important problem at this time is the development of a renormalizable unified theory of all the interactions. We have seen that we can cope with the incomplete unification in the Standard Model between the electroweak interactions and the strong interaction. The Standard Model is renormalizable and thus we can make perturbation theory calculations with confidence and compare the results with experiment. But we cannot make calculations in quantum gravity with confidence. As a result Planck scale physics is totally speculative and we cannot understand the nature of the Big Bang when the universe was contained within a region the size of the Planck length.

With these issues in mind we have developed a unified theory of all the known interactions that is finite. Thus we can perform calculations to any order of perturbation theory in any sector and obtain finite results that can be compared with experiment. The theory has the Standard Model and classical General Relativity as "low energy" limits. At high energies a

[3] This historical process of Physics is described in some detail in the book *Cosmos and Consciousness* by this author.

string-like sub-structure generates a smooth high-energy limit that eliminates the divergences in perturbation theory.

The nature of the string-like sub-structure is not dependent on the details of the interactions and symmetries. The interactions in the Standard Model and quantum gravity are can have any polynomial form (in the quantum fields). A variety of other interactions (such as those with derivative couplings) are allowed in this approach which do not affect the finiteness of the theory. Thus the range of possible extensions of the Standard Model is significantly widened.

2. An Extension of the Calculus of Variations

A New Paradigm in the Calculus of Variations

The Calculus of Variations has a long and venerable history in Physics and Mathematics. Many problems in Physics and Mathematics have been treated with approaches based on techniques in the Calculus of Variations (see References 1 – 6 at the end of the book). In this book we will develop a unified quantum field theory of the known forces of nature based on a new type of problem, or paradigm, in the Calculus of Variations. One way of viewing the spectrum of problems in the calculus of variations is the following progression.

A Classification of Variational Problems

1. Variational problems in a Euclidean, or Minkowski, flat space such as the minimal distance between two points or the extrema of a field theory Lagrangian.

2. Variational problems seeking extrema on a curved surface such as the shortest distance between points on the surface of a sphere.

This book develops a third and fourth, possibility, that to the author's knowledge, has not been addressed in the literature:

3. Variational problems where the extrema are determined on a surface that is itself defined as an extremum. The discussions in this book exemplify this pardigm.

4. Variational problems where the extrema are determined on a surface that is itself defined as an extremum that depends on the extrema on the surface. More simply put the extrema, and the surface upon which they are defined, are jointly determined and are interrelated. Fortunately, our unified theory does not use this paradigm. A future theory might.

In the unified theory that we will develop all particle fields including the graviton field are defined as a mapping of space-time into itself with the mapping determined as an extremum of a variation of a fundamental field (a type 3 variational problem in the above classification). Our theory could be generalized to include a back-reaction of the particle fields on the fundamental field (a type 4 variational problem in the above classification). We will not discuss this possibility in this book.

Simple Physical Example – Strings On Springs

In this section we will describe a simple physical example that illustrates a variational problem of type 3 in the Calculus of Variations. We view it as a composition of extrema. The approach used in the solution of this problem will be used in the development of the unified quantum field theory subsequently.

A Strings on Springs Mechanics Problem

Consider a long string or bar that can oscillate (undulate) in a direction perpendicular to its length. Further assume that one end of this bar or string is attached to a spring that cause the entire bar or string to oscillate back and forth in a direction parallel to its long side. This configuration is illustrated in Fig. 2.1.

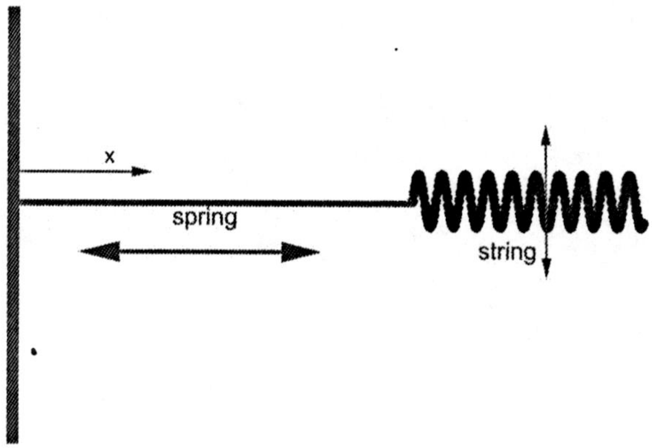

Figure 2.1. An oscillating string attached to a spring.

Let x denote the distance to a point on the string when the spring is at equilibrium. If 2π times the frequency of the spring is ω_1, then the location of this point when the spring is oscillating is

$$X(t) = x + A \sin(\omega_1 t + \phi_1) \tag{2.1}$$

where ϕ_1 is a phase, and A is the amplitude of the spring oscillation. The vertical displacement of a traveling wave on the string can take the form

$$\psi(t) = B \sin(\omega_2 t - k_2(x + A \sin(\omega_1 t + \phi_1)) + \phi_2) \tag{2.2}$$

where B is the amplitude of the string wave, and k_2, ω_2 and ϕ_2 are the parameters of the string wave. These simple mechanical formulae are well known. But they lead to an interesting new application of the ideas of the Calculus of Variations.

Suppose we treat X as an independent variable with X given by eq. (2.1), and with eq. (2.2) written as:

$$\psi(t) = B \sin(\omega_2 t - k_2 X + \phi_2) \tag{2.3}$$

Defining

$$\psi = \psi(X(t), t) \tag{2.4}$$

we can specify the dynamics of the above motion by finding the extrema of

$$I = \int \mathscr{L}_\psi \, dX(t) + \int \mathscr{L}_X \, dt \tag{2.5}$$

where the Lagrangian terms are

$$\mathscr{L}_\psi = \tfrac{1}{2} \{ \mu \, (\partial\psi/\partial t)^2 - Y \, (\partial\psi/\partial X)^2 \} \tag{2.6}$$

with μ and Y being constants, and

$$\mathscr{L}_X = \tfrac{1}{2} \{ m(\partial X/\partial t)^2 - k(X - x)^2 \} \tag{2.7}$$

where m and k are constants, and where x is a parameter. Applying Hamilton's Principle, and performing independent variations of X and ψ yields the Lagrangian equations:

$$\frac{\partial \mathscr{L}_\psi}{\partial \psi} - \frac{\partial}{\partial X} \frac{\partial \mathscr{L}_\psi}{\partial(\partial\psi/\partial X)} - \frac{\partial}{\partial t} \frac{\partial \mathscr{L}_\psi}{\partial(\partial\psi/\partial t)} = 0 \tag{2.8}$$

and

$$\frac{\partial \mathscr{L}_X}{\partial X} - \frac{\partial}{\partial t} \frac{\partial \mathscr{L}_X}{\partial(\partial X/\partial t)} = 0 \tag{2.9}$$

The resulting equations of motion are:

$$\mu\, \partial^2\psi/\partial t^2 - Y\, \partial^2\psi/\partial X^2 = 0 \qquad (2.10)$$

and

$$m\, \partial^2 X/\partial t^2 + k(X - x) = 0 \qquad (2.11)$$

with the solutions given in eqs. 2.1 and 2.2.

The procedure that we use to obtain these results may look a bit strange but they illustrate a type 3 problem in the Calculus of Variations involving the composition of extrema—the composition of an extremum that specifies a manifold in a space (possibly including all of space in a $R^n \rightarrow R^n$ mapping) with an extremum determining a function on that manifold. The procedure is described in detail in the next section.

The Composition of Extrema – Lagrangian Formulation

In this section we will explore the general case of the composition of extrema for fields. We will discuss the case of a scalar field ϕ which is a function of a vector field X^μ in a D-dimensional space with coordinate variables that we will denote as y^μ. (The discussion for other types of fields is a straightforward extension of this discussion.) Thus

$$\phi = \phi(X) \qquad (2.12)$$

and

$$X^\mu = X^\mu(y) \qquad (2.13)$$

We assume that the dynamics can be described by a Lagrangian formulation using an extension of Hamilton's principle:

$$I = \int \mathscr{L} d^4y \qquad (2.14)$$

with

$$\mathscr{L} = \mathscr{L}(\phi(X), \partial\phi/\partial X^\nu, X^\mu(y), \partial X^\mu(y)/\partial y^\nu, y) \qquad (2.15)$$

If we perform a standard variation[4] in ϕ for fixed y (and thus fixed X) we find

---

[4] Bogoliubov, N. N., & Shirkov, D. V., Volkoff, G. M. (tr), *Introduction to the Theory of Quantized Fields* (Wiley-Interscience, New York, 1959); Goldstein H., *Classical Mechanics* (Addison-Wesley, Reading, MA 1965).

$$\delta I = \int [\delta\phi\, \partial\mathscr{L}/\partial\phi \;+ \delta(\partial\phi/\partial X^\nu)\, \partial\mathscr{L}/\partial(\partial\phi/\partial X^\nu)]\, d^4y \qquad (2.16)$$

We can rewrite the variation in the derivative of ϕ as

$$\delta(\partial\phi/\partial X^\nu) = \partial(\delta\phi)/\partial X^\nu \qquad (2.17)$$

$$= \partial y^\mu/\partial X^\nu\, \partial(\delta\phi)/\partial y^\mu \qquad (2.18)$$

with an implied summation over repeated indices. After substituting eq. 2.18 in eq. 2.16, and performing an integration by parts (and discarding the surface term which is assumed to yield zero in the standard fashion) we obtain:

$$\delta I = \int \delta\phi\, \{\partial\mathscr{L}/\partial\phi \;- \partial/\partial y^\mu\, [\partial\mathscr{L}/\partial(\partial\phi/\partial X^\nu)\, \partial y^\mu/\partial X^\nu)\,]\}\, d^4y$$

Since the variation of $\delta\phi$ is arbitrary we conclude

$$\partial\mathscr{L}/\partial\phi \;- \partial/\partial y^\mu\, [\partial\mathscr{L}/\partial(\partial\phi/\partial X^\nu)\, \partial y^\mu/\partial X^\nu)] = 0 \qquad (2.19a)$$

Similarly we can perform a variation in X^μ and obtain

$$\partial\mathscr{L}/\partial X^\mu \;- \partial/\partial y^\nu\, [\partial\mathscr{L}/\partial(\partial X^\mu/\partial y^\nu)] = 0 \qquad (2.19b)$$

The second term in eq. 2.19a shows the effect of the dependence of ϕ on the field X, $\phi = \phi(X)$, rather than directly on the coordinate system y. The X field defines a "manifold" or, more properly, specifies a transformation from $R^n \rightarrow R^n$. If we make standard assumptions about the mapping: that it is continuous and piece-wise invertible, then we can establish the following lemmas:

Lemma 1: *If the transformation $X^\mu = X^\mu(y)$ is a transformation from $R^n \rightarrow R^n$ that is of class C' and piece-wise invertible, then*

$$\frac{\partial}{\partial y^\nu}\frac{\partial y^\nu}{\partial X^\mu} = -\frac{\partial}{\partial X^\mu}\ln J \qquad (2.20)$$

where

$$J = |\partial(X)/\partial(y)| \tag{2.21}$$

is the absolute value of the Jacobian of the transformation.

Proof:
Consider two equivalent forms of an integral:

$$I = \int \mathscr{L} J\, d^4y = \int \mathscr{L}\, d^4X$$

where $\mathscr{L}$ is specified as in eq. 2.15. Then the first expression for I leads to eq. 2.19a which can be written in the form

$$\partial\mathscr{L}/\partial\phi \;-\; \partial/\partial X^\mu\, [\partial\mathscr{L}/\partial(\partial\phi/\partial X^\mu)] - \partial\mathscr{L}/\partial(\partial\phi/\partial X^\mu)\, \{\partial[J\, \partial y^\nu/\partial X^\mu]/\partial y^\nu\} = 0$$

Using the second expression for I above we obtain the following equation by variation in ϕ:

$$\partial\mathscr{L}/\partial\phi \;-\; \partial/\partial X^\mu\, [\partial\mathscr{L}/\partial(\partial\phi/\partial X^\mu)] = 0$$

Comparing these two expressions and realizing that $\partial\, [J\partial y^\nu/\partial X^\mu]/\partial y^\nu$ is totally independent of ϕ and its derivatives leads us to conclude

$$\partial\, [J\partial y^\nu/\partial X^\mu]/\partial y^\nu \;= 0 \tag{2.22}$$

It is a general relationship for a transformation between X and y based on continuity and piece-wise invertibility. After a few elementary manipulations eq. 2.22 can be rewritten in the form of eq. 2.20. ■

Lemma 2: *If the transformation $X^\mu = X^\mu(y)$ is a transformation from $R^n \to R^n$ that is of class C′ and piece-wise invertible and* $\mathscr{L} = \mathscr{L}(\phi(X), \partial\phi/\partial X^\nu, X^\mu(y), \partial X^\mu(y)/\partial y^\nu, y)$*, then*

$$\partial\mathscr{L}/\partial(\partial\phi/\partial X^\nu)\, \partial y^\mu/\partial X^\nu = \partial\mathscr{L}/\partial(\partial\phi/\partial y^\mu) \tag{2.23}$$

Proof:

Let us express $\mathscr{L}$ as a power series in derivatives of ϕ:

$$\mathscr{L} = \sum_{n=0} a_{n\mu_1\mu_2\cdots\mu_n}(\phi(X), X^\mu(y), \partial X^\mu(y)/\partial y^\nu, y) \prod_{j=1}^{n} \partial\phi/\partial X^{\mu_j}$$

which can rewritten using piece-wise invertibility as

$$\mathscr{L} = \sum_{n=0} a_{n\mu_1\mu_2\cdots\mu_n}(\phi(X), X^\mu(y), \partial X^\mu(y)/\partial y^\nu, y) \prod_{j=1}^{n} \partial\phi/\partial y^{\nu_j}\, \partial y^{\nu_j}/\partial X^{\mu_j}$$

Taking the derivative of this equation with respect to $\partial\phi/\partial y^\mu$ immediately yields the result. ■

Eq. 2.23 enables us to rewrite eq. 2.19a as:

$$\partial\mathscr{L}/\partial\phi - \partial/\partial y^\mu\, [\partial\mathscr{L}/\partial(\partial\phi/\partial y^\mu)] = 0 \tag{2.24}$$

which is as one would expect.

In order to get a feeling for the effect of eq. 2.19a we will look at a simple example where we specify the relation of the X and y variables directly. Then we will look at the composition of extrema where the transformation between X and y is itself determined as an extremum solution.

Example: a hyperplane

We assume eq. 2.19b yields the transformation:

$$X^i = ay^i \qquad \text{for } i = 1,2,3$$

$$X^0 = 0$$

Then eq. 2.19a becomes

$$\partial\mathscr{L}/\partial\phi - \partial/\partial y^i\, [\partial\mathscr{L}/\partial(\partial\phi/\partial y^i)] = 0 \tag{2.25}$$

with the time derivative disappearing. Effectively the variation of ϕ on the hyperplane $X^0 = 0$ is determined by the differential equation generated by 2.25. On this hyperplane the transformation between the X and y variables is invertible.

Coordinate Transformation Determined as an Extremum Solution

We now develop a formalism that determines a mapping from space onto itself as the solution of an extremum problem and also determines the dynamics of one or more fields as a function of this mapping. To this author's knowledge this area in the Calculus of Variations – the determination of an extremum on a manifold where the manifold itself is determined by an extremum – has not been previously explored. We will also develop a hamiltonian formulation. Then we will proceed to quantize the theory.

Separable Lagrangian Case

Although there are many forms that the composition of extrema could take, one fairly general form that is directly useful in quantum field theory applications is based on a Lagrangian that can be split into two parts which we will call a *separable Lagrangian*:

$$\mathscr{L} = \mathscr{L}_F\, J + \mathscr{L}_C(X^\mu(y), \partial X^\mu(y)/\partial y^\nu, y) \tag{2.26}$$

where J is defined in eq. 2.21, where $\mathscr{L}_F$ contains all the dynamics of the fields and their interactions, and where $\mathscr{L}_C$ defines the coordinate mapping as an extremum solution. The procedure to determine the differential equations that specify the mapping, and the field equations that specify field interactions and evolution, is to vary in the coordinates X^μ and in the fields independently, using Hamilton's Principle. The extrema are to be determined for

$$I = \int \mathscr{L}\, d^4y \tag{2.27}$$

We will begin by considering the case of one scalar field:

$$\mathscr{L}_F = \mathscr{L}_F(\phi(X), \partial\phi/\partial X^\nu) \tag{2.28}$$

and

$$\mathscr{L}_C = \mathscr{L}_C(X^\mu(y), \partial X^\mu(y)/\partial y^\nu, y) \tag{2.29}$$

Eq. 2.27 can be written in the form:

$$I = \int \mathscr{L}_F(\phi(X), \partial\phi/\partial X^\nu)\, dX + \int \mathscr{L}_C(X^\mu(y), \partial X^\mu(y)/\partial y^\nu, y)\, d^4y \tag{2.30}$$

using the Jacobian to transform to an integral over dX in the first term. A standard variation of ϕ and the application of Hamilton's Principle yields

$$\partial\mathscr{L}_F/\partial\phi - \partial/\partial X^\mu\, [\partial\mathscr{L}_F/\partial(\partial\phi/\partial X^\mu)] = 0 \tag{2.31}$$

reflecting the fact that ϕ is a function of X^μ only, with X^μ a function of the y coordinates.

Next we perform a variation of X^μ determining the mapping from $y \rightarrow X$ as an extremum of the integral in eq. 2.27. We note the piece-wise invertibility of the coordinate mapping $X^\mu(y)$ allows us to write the Jacobian J as a function of y^μ only. A standard variation of X^μ and the application of Hamilton's Principle yields

$$\partial\mathscr{L}_C/\partial X^\mu - \partial/\partial y^\nu\,[\partial\mathscr{L}_C/\partial(\partial X^\mu/\partial y^\nu)] = 0 \qquad (2.32)$$

Klein-Gordon Example

The Klein-Gordon scalar field theory furnishes us with a simple example of the application of the preceding development. The Lagrangian is

$$\mathscr{L}_F = ½\,[\,(\partial\phi/\partial X^\nu)^2 - m^2\phi^2\,] \qquad (2.33)$$

From eq. 2.31 we obtain the field equation:

$$(\Box + m^2)\,\phi(X) = 0 \qquad (2.34)$$

where

$$\Box = \partial/\partial X^\nu\,\partial/\partial X_\nu \qquad (2.34a)$$

A fourier representation of the solution of eq. 2.34 is:

$$\phi(X) = \int dp\,\delta(p^2 - m^2)\theta(p^0)\,[A(p)\,e^{-ip.X} + A(p)^*\,e^{ip.X}] \qquad (2.35)$$

where A(k) is a function of k and * indicates complex conjugation.

The determination of $X^\mu(y)$ depends on the Lagrangian $\mathscr{L}_C$ and the solutions of eq. 2.32. If we chose

$$\mathscr{L}_C = -½\,(\partial X^\mu/\partial y^\nu)^2 \qquad (2.36)$$

and obtain the equation

$$\Box\,X^\mu = 0 \qquad (2.37)$$

with the solution

$$X^\mu = \int dk\, \delta(k^2)\theta(k^0)\, [a^\mu(k)\, e^{-ik.y} + a^\mu(k)^*\, e^{ik.y}] \qquad (2.38)$$

where $a^\mu(k)$ are complex vector functions of k in general. (We ignore positivity issues for the moment.) Substitution of eq. 2.38 in eq. 2.35 yields an expression with a form reminiscent of bosonic string expressions.[5] We will take up this point later in subsequent chapters.

The Composition of Extrema – Hamiltonian Formulation

The previous section established a Lagrangian formulation of dynamics based on the composition of extrema. In this section we will develop an equivalent hamiltonian formulation. We will assume a Minkowskian space-time with X^0 and y^0 playing the role of the time coordinates in the respective coordinate systems.

Initially, we will assume a scalar field ϕ with a Lagrangian of the form in eq. 2.15 and define canonical momenta with

$$\Pi_\phi = \partial\mathscr{L}/\partial\dot{\phi} \equiv \partial\mathscr{L}/\partial(\partial\phi/\partial X^\mu)\, \partial y^0/\partial X^\mu \qquad (2.39)$$

$$\Pi_X{}^\mu = \partial\mathscr{L}/\partial\dot{X}_\mu \qquad (2.40)$$

where

$$\dot{\phi} = \partial\phi/\partial y^0 \equiv \partial\phi/\partial X^\mu\, \partial X^\mu/\partial y^0 \qquad (2.41)$$

$$\dot{X}^\mu = \partial X^\mu/\partial y^0 \qquad (2.42)$$

The we define the hamiltonian density as

$$\mathscr{H} = \Pi_\phi\, \dot{\phi} + \Pi_X{}^\mu\, \dot{X}_\mu - \mathscr{L}(\phi(X),\, \partial\phi/\partial X^\nu,\, X^\mu(y),\, \partial X^\mu(y)/\partial y^\nu,\, y) \qquad (2.43)$$

and the hamiltonian

$$H = \int \mathscr{H}\, d^3y \qquad (2.44)$$

where the integration over y does *not* include the time coordinate y^0.

The hamiltonian density has the general form

---

[5] See for example Joseph Polchinski, *String Theory* (Cambridge University Press, New York, 1998). D. Bailin and A. Love, *Supersymmetric Gauge Field Theory and String Theory* (Institute of Physics Publishing, Philadelphia, PA, 1994).

$$\mathscr{H} = \mathscr{H}(\ \phi(X), \partial\phi/\partial X^i, \Pi_\phi, X^\mu(y), \partial X^\mu(y)/\partial y^j, \Pi_X{}^\mu, y^\nu) \qquad (2.45)$$

for the case of one scalar field where the indices i and j represent space coordinates; time coordinates are assigned index value 0.

If we calculate the differential change in H using eq. 2.45 we obtain

$$dH = \int\{\partial\mathscr{H}/\partial\phi\ d\phi + \partial\mathscr{H}/\partial\Pi_\phi\ d\Pi_\phi - \partial/\partial y^\nu\ [\partial\mathscr{H}/\partial(\partial\phi/\partial X^j)\ \partial y^\nu/\partial X^j]\ d\phi +$$

$$+ \partial\mathscr{H}/\partial X^\mu\ dX^\mu + \partial\mathscr{H}/\partial\Pi_X{}^\mu\ d\Pi_X{}^\mu - \partial/\partial y^j\ [\partial\mathscr{H}/\partial(\partial X^\mu/\partial y^j)]\ dX^\mu\}d^3y \qquad (2.46)$$

after some partial integrations. (Repeated indices indicate summations. Indices labeled i and j indicate space coordinates. Greek indices include all space-time components of a variable.)

Expressing the differential in H using eq. 2.43 we obtain

$$dH = \int dy\ \{\Pi_\phi\ d\dot\phi + \dot\phi\ d\Pi_\phi - \partial\mathscr{L}/\partial\phi\ d\phi - \partial\mathscr{L}/\partial(\partial\phi/\partial X^\mu)d(\partial\phi/\partial X^\mu) +$$

$$+ \Pi_X{}^\mu\ d\dot X^\mu + \dot X^\mu\ d\Pi_X{}^\mu - \partial\mathscr{L}/\partial X^\mu\ dX^\mu - \partial\mathscr{L}/\partial(\partial X^\mu/\partial y^j)\ d(\partial X^\mu/\partial y^j)\} \qquad (2.47a)$$

After some manipulations we find

$$dH = \int\{\ \dot\phi\ d\Pi_\phi + \dot X_\mu d\Pi_X{}^\mu\ - \partial/\partial y^0\ \Pi_\phi\ d\phi - \partial/\partial y^0\ \Pi_X{}^\mu\ dX_\mu\}\ dy \qquad (2.47b)$$

using the equations of motion eqs. 2.19a and 2.19b.

Comparing eqs 2.46 and 2.47 we obtain Hamilton's equations in the case of the composition of extrema:

$$\dot\phi\ = \partial\mathscr{H}/\partial\Pi_\phi \qquad (2.48a)$$

$$\dot\Pi_\phi = -\partial\mathscr{H}/\partial\phi + \partial/\partial y^\nu\ [\partial\mathscr{H}/\partial(\partial\phi/\partial X^j)\ \partial y^\nu/\partial X^j] \qquad (2.48b)$$

$$\dot X_\mu\ = \partial\mathscr{H}/\partial\Pi_X{}^\mu \qquad (2.48c)$$

$$\dot\Pi_X{}^\mu = -\partial\mathscr{H}/\partial X^\mu + \partial/\partial y^j\ [\partial\mathscr{H}/\partial(\partial X^\mu/\partial y^j)] \qquad (2.49d)$$

where

$$\dot{\Pi}_\phi = \partial\, \Pi_\phi/\partial y^0 \quad (2.49a)$$

$$\dot{\Pi}_{X^\mu} = \partial \Pi_{X^\mu}/\partial y^0 \quad (2.49b)$$

Translational Invariance

If the Lagrangian of a field theory has no explicit dependence on the coordinates then one expects translational invariance accompanied by a conservation law for an energy-momentum stress tensor. We will show this is the case for Lagrangians implementing the composition of extrema. We assume a Lagrangian without an explicit dependence on the coordinates y^ν:

$$\mathscr{L} = \mathscr{L}(\phi(X), \partial\phi/\partial X^\nu, X^\mu(y), \partial X^\mu(y)/\partial y^\nu) \quad (2.50)$$

Under an infinitesimal displacement,

$$y'^\nu = y^\nu + \epsilon^\nu \quad (2.51a)$$

$$\delta\phi = \phi(X(y + \epsilon)) - \phi(X(y))$$

$$= \epsilon^a\, \partial\phi/\partial y^a \quad (2.51b)$$

$$\delta X^\mu = \epsilon^a\, \partial X^\mu/\partial y^a \quad (2.51c)$$

$$\delta(\partial\phi/\partial X^\mu) = \epsilon^a\, \partial(\partial\phi/\partial y^a)/\partial X^\mu \quad (2.51d)$$

$$\delta(\partial X^\mu/\partial y^\nu) = \epsilon^a\, \partial(\partial X^\mu/\partial y^a)/\partial y^\nu \quad (2.51e)$$

and the Lagrangian changes by

$$\delta\mathscr{L} = \epsilon^a\, \partial\mathscr{L}/\partial y^a \quad (2.52)$$

The change can also be expressed in terms of the changes in the fields, their derivatives and the mapping X^μ:

$$\delta\mathcal{L} = \partial\mathcal{L}/\partial\phi\ \delta\phi + \partial\mathcal{L}/\partial(\partial\phi/\partial X^\mu)\ \delta(\partial\phi/\partial X^\mu) + \partial\mathcal{L}/\partial X^\mu\ \delta X^\mu + \\ + \partial\mathcal{L}/\partial(\partial X^\mu/\partial y^\nu)\ \delta(\partial X^\mu/\partial y^\nu) \tag{2.53}$$

Combining eqs. 2.51, 2.52 and 2.53 we obtain (after some manipulations):

$$\in^\nu \partial/\partial y_\mu\ \mathcal{T}_{\mu\nu} = 0 \tag{2.54}$$

where

$$\mathcal{T}_{\mu\nu} = -\ g_{\mu\nu}\mathcal{L} + \partial\mathcal{L}/\partial(\partial\phi/\partial X^\delta)\ \partial y_\mu/\partial X^\delta\ \partial\phi/\partial y^\nu + \partial\mathcal{L}/\partial(\partial X^\delta/\partial y_\mu)\ \partial X^\delta/\partial y^\nu \tag{2.55a}$$

or, alternately using Lemma 2,

$$\mathcal{T}_{\mu\nu} = -\ g_{\mu\nu}\mathcal{L} + \partial\mathcal{L}/\partial(\partial\phi/\partial y_\mu)\ \partial\phi/\partial y^\nu + \partial\mathcal{L}/\partial(\partial X^\delta/\partial y_\mu)\ \partial X^\delta/\partial y^\nu \tag{2.55b}$$

Since $\in^a$ is an arbitrary displacement we obtain the conservation law:

$$\partial/\partial y_\mu\ \mathcal{T}_{\mu\nu} = 0 \tag{2.56}$$

Eq. 2.56 implies the energy-momentum vector

$$P_\beta = \int d^3y\ \mathcal{T}_{0\beta} \tag{2.57}$$

is conserved. We note

$$\partial/\partial y^0\ P_\beta = 0 \tag{2.58}$$

since eq. 2.56 and 2.57 can be used to obtain the integral of a divergence, which results in zero.
The hamiltonian (eqs. 2.43-44) is

$$H = P_0 \tag{2.59}$$

We note for later use that the total energy, H, which is conserved, contains a term that represents the energy in the X^μ mapping. Thus energy can be exchanged in principle between the ϕ field sector and the X^μ sector.

Lorentz Invariance and Angular Momentum Conservation

We can also verify Lorentz invariance and obtain the form of the conserved angular momentum by considering the effect of an infinitesimal Lorentz transformation. We will consider the case of a scalar field ϕ.

Under an infinitesimal Lorentz transformation ($\epsilon_{\mu\nu} = -\epsilon_{\nu\mu}$):

$$y'_\mu = y_\mu + \delta y_\mu = y_\mu + \epsilon_{\mu\nu} y^\nu \tag{2.60a}$$

$$\delta\phi = \phi(X(y')) - \phi(X(y))$$

$$= \epsilon^{\mu\nu}\, y_\nu\, \partial\phi/\partial X^a\, \partial X^a/\partial y^\mu \tag{2.60b}$$

$$\delta X^\mu = S^\mu{}_a X^a(y') - X^\mu(y) \tag{2.60c}$$

$$= \epsilon^\mu{}_a X^a(y) + \partial X^\mu/\partial y^\beta\, \delta y^\beta \tag{2.60d}$$

where $S^\mu{}_a$ is the matrix for the Lorentz transformation of a vector. (If X^μ were a gauge field then an additional operator gauge term would have to be added to eq. 2.60d.)

The Lagrangian changes by

$$\delta\mathscr{L} = \epsilon^{\mu\nu}\, y_\nu\, \partial\mathscr{L}/\partial y^\mu \tag{2.61}$$

under the infinitesimal Lorentz transformation. The change in the Lagrangian can also be expressed as:

$$\delta\mathscr{L} = \partial\mathscr{L}/\partial\phi\, \delta\phi + \partial\mathscr{L}/\partial(\partial\phi/\partial X^\mu)\, \delta(\partial\phi/\partial X^\mu) + \partial\mathscr{L}/\partial X^\mu\, \delta X^\mu + {} + \partial\mathscr{L}/\partial(\partial X^\mu/\partial y^\nu)\, \delta(\partial X^\mu/\partial y^\nu) \tag{2.62}$$

Combining eqs. 2.61 and 2.62, and substituting and simplifying terms leads to:

$$\epsilon_{\mu\nu}\, \partial/\partial y^\sigma\, \mathscr{M}^{\sigma\mu\nu} = 0 \tag{2.63}$$

where

$$\mathscr{M}^{\sigma\mu\nu} = (g^{\mu\sigma} y^\nu - g^{\nu\sigma} y^\mu)\mathscr{L} + \partial\mathscr{L}/\partial(\partial\phi/\partial X^a)\, \partial y^\sigma/\partial X^a\, (y^\mu \partial\phi/\partial y_\nu - y^\nu \partial\phi/\partial y_\mu) + {} + \partial\mathscr{L}/\partial(\partial X^\delta/\partial y^\sigma)\, (g^{\delta\nu} X^\mu - g^{\delta\mu} X^\nu + y^\mu\, \partial X^\delta/\partial y_\nu - y^\nu\, \partial X^\delta/\partial y_\mu) \tag{2.64}$$

The conserved angular momentum is:

$$M^{\mu\nu} = \int d^3y\, \mathcal{M}^{0\mu\nu} \tag{2.65}$$

with

$$\partial M^{\mu\nu}/\partial y^0 = 0 \tag{2.66}$$

The angular momentum density can be written in the familiar form:

$$\mathcal{M}^{\sigma\mu\nu} = y^\mu\, \mathcal{T}^{\sigma\nu} - y^\nu\, \mathcal{T}^{\sigma\mu} + \partial\mathcal{L}/\partial(\partial X^\delta/\partial y^\sigma)\, (g^{\delta\nu}X^\mu - g^{\delta\mu}X^\nu) \tag{2.67}$$

taking account of the vector nature of X^μ. The spatial part of $M^{\mu\nu}$ is the angular momentum.

Internal Symmetries

We will now consider the case of a set of scalar fields ϕ_r in a Lagrangian with an internal symmetry. Under a local transformation

$$\phi_r(X) \rightarrow \phi_r(X) - i\epsilon\lambda_{rs}\, \phi_s(X) \tag{2.68}$$

If the Lagrangian is invariant under this transformation, then

$$\delta\mathcal{L} = 0 = \partial\mathcal{L}/\partial\phi_r\, \delta\phi_r + \partial\mathcal{L}/\partial(\partial\phi_r/\partial X^\alpha)\, \delta(\partial\phi_r/\partial X^\alpha) \tag{2.69}$$

Using the equation of motion eq. 2.19a satisfied by all the components ϕ_r we obtain a conserved current:

$$\mathcal{J}^\nu = -i\, \partial\mathcal{L}/\partial(\partial\phi_r/\partial X^\delta)\, \partial y^\nu/\partial X^\delta\, \lambda_{rs}\, \phi_s \tag{2.70}$$

which satisfies

$$\partial\mathcal{J}^\nu/\partial y^\nu = 0 \tag{2.71}$$

The conserved charge is

$$Q = \int d^3y\, \mathcal{J}^0 \tag{2.72}$$

$$\partial Q/\partial y^0 = 0 \tag{2.73}$$

where the integration in eq, 2.72 is over the space components of y only.

Separable Lagrangians

We now consider the case of a separable Lagrangian such as in eq. 2.26. Adopting the definitions:

$$\phi' = \partial\phi/\partial X^0 \tag{2.74}$$

$$X_\mu' = \partial X_\mu/\partial y^0 \tag{2.75}$$

we define canonical momenta as

$$\pi_\phi = \partial\mathscr{L}/\partial\phi' \equiv \partial\mathscr{L}/\partial(\partial\phi/\partial X^0) \tag{2.76}$$

$$\pi_X{}^\mu = \partial\mathscr{L}/\partial X_\mu' \equiv \partial\mathscr{L}/\partial(\partial X_\mu/\partial y^0) \tag{2.77}$$

We now define the separable hamiltonian density as

$$\mathscr{H}_s = J\pi_\phi\,\phi' + \pi_X{}^\mu\, X_\mu' - \mathscr{L}_s \tag{2.78}$$

where J is the Jacobian (eq. 2.21) and

$$H_s = \int \mathscr{H}_s\, d^3y \tag{2.79}$$

The separable Lagrangian (from eq. 2.26) is:

$$\mathscr{L}_s = \mathscr{L}_F(\phi(X), \partial\phi/\partial X^\mu)\, J + \mathscr{L}_C(X^\mu(y), \partial X^\mu(y)/\partial y^\nu, y) \tag{2.80}$$

In the case of one scalar field the separable hamiltonian density has the general form

$$\mathscr{H}_s = \mathscr{H}_s(\, \phi(X), \pi_\phi, \partial\phi/\partial X^i, X^\mu(y), \pi_X{}^\mu, \partial X^\mu(y)/\partial y^j, y^\nu) \tag{2.81}$$

where the indices i and j indicate spatial components. In particular, the terms in the separable hamiltonian are:

$$\mathscr{H}_S = \mathscr{H}_F\, J + \mathscr{H}_C \tag{2.82}$$

with

$$\mathscr{H}_F(\phi(X), \pi_\phi, \partial\phi/\partial X^i) = \pi_\phi\, \phi' - \mathscr{L}_F \tag{2.83}$$

$$\mathscr{H}_C(X^\mu(y), \pi_X{}^\mu, \partial X^\mu(y)/\partial y^j, y^\nu) = \pi_X{}^\mu\, X_\mu{}' - \mathscr{L}_C \tag{2.84}$$

where J is the absolute value of the Jacobian defined in 2.21.

We now define the time integral of H as we did in eq. 2.14 when considering the Lagrangian formulation:

$$G = \int dy^0\, H_s \tag{2.85}$$

Thus G is an integral over all space-time coordinates. Using G we can develop a hamiltonian formulation. First we calculate the differential change in G. Using eqs. 2.81-2 and 2.85 we obtain

$$\begin{aligned} dG = \int\Big\{ & J\, \partial\mathscr{H}_F/\partial\phi\, d\phi + J\, \partial\mathscr{H}_F/\partial\pi_\phi\, d\pi_\phi + \\ & + J\, \partial\mathscr{H}_F/\partial(\partial\phi/\partial X^j)\, d(\partial\phi/\partial X^j) + \partial\mathscr{H}_C/\partial X^\mu\, dX^\mu + \\ & + \partial\mathscr{H}_C/\partial\pi_X{}^\mu\, d\pi_X{}^\mu + \partial\mathscr{H}_C/\partial(\partial X^\mu/\partial y^j)\, d(\partial X^\mu/\partial y^j) \Big\}\, d^4y \end{aligned} \tag{2.86}$$

with summations implied by repeated indices. (Index labels i and j label spatial coordinates only; Greek indices label space-time coordinates.) Rewriting dG as two integrals and performing partial integrations yields:

$$\begin{aligned} dG = & \int d^4X \Big\{\partial\mathscr{H}_F/\partial\phi\, d\phi + \partial\mathscr{H}_F/\partial\pi_\phi\, d\pi_\phi - \partial/\partial X^j[\partial\mathscr{H}_F/\partial(\partial\phi/\partial X^j)]\, d\phi \Big\} + \\ & + \int d^4y \Big\{\partial\mathscr{H}_C/\partial X^\mu\, dX^\mu + \partial\mathscr{H}_C/\partial\pi_X{}^\mu\, d\pi_X{}^\mu - \partial/\partial y^j[\partial\mathscr{H}_C/\partial(\partial X^\mu/\partial y^j)]\, dX^\mu \Big\} \end{aligned} \tag{2.87}$$

Alternately, expressing the differential in G using eqs. 2.82-4 we obtain

$$\begin{aligned} dG = & \int d^4X \Big\{\pi_\phi\, d\phi' + \phi'\, d\pi_\phi - \partial\mathscr{L}_F/\partial\phi\, d\phi - \partial\mathscr{L}_F/\partial(\partial\phi/\partial X^\mu) d(\partial\phi/\partial X^\mu)\Big\} + \\ & + \int d^4y \Big\{\pi_{X\mu}\, dX^{\mu\prime} + X^{\mu\prime}\, d\pi_{X\mu} - \partial\mathscr{L}_C/\partial X^\mu\, dX^\mu - \partial\mathscr{L}_C/\partial(\partial X^\mu/\partial y^j)\, d(\partial X^\mu/\partial y^j)\Big\} \end{aligned} \tag{2.88}$$

which becomes

$$dG = \int d^4X \left\{-\pi_\phi{}' \, d\phi + \phi' \, d\pi_\phi\right\} + \int d^4y \left\{-\pi_{X\mu}{}' \, dX^\mu + X^{\mu\prime} \, d\pi_{X\mu}\right\} \quad (2.89)$$

using the equations of motion eqs. 2.31-2.

Comparing eqs 2.87 and 2.89 we obtain Hamilton's equations for the case of the composition of extrema for a separable Lagrangian:

$$\phi' = \partial\mathscr{H}_F/\partial\pi_\phi \quad (2.90)$$

$$\pi_\phi{}' = -\,\partial\mathscr{H}_F/\partial\phi + \partial/\partial X^j\,[\partial\mathscr{H}_F/\partial(\partial\phi/\partial X^j)] \quad (2.91)$$

$$X_\mu{}' = \partial\mathscr{H}_C/\partial\pi_X{}^\mu \quad (2.92)$$

$$\pi_{X\mu}{}' = -\,\partial\mathscr{H}_C/\partial X^\mu + \partial/\partial y^j\,[\partial\mathscr{H}_C/\partial(\partial X^\mu/\partial y^j)] \quad (2.93)$$

where

$$\pi_\phi{}' = \partial\,\pi_\phi/\partial X^0 \quad (2.94)$$

$$\pi_{X\mu}{}' = \partial\,\pi_{X\mu}/\partial X^0 \quad (2.95)$$

Notice that $\mathscr{L}_F$, $\mathscr{H}_F$ *and* π_ϕ *have precisely the same form, as a function of* X^μ, *as one sees in a conventional field theory formalism.* Yet X^μ is a mapping/function of the coordinates y. In reality, it can be viewed as a field as we shall see.

Separable Lagrangians and Translational Invariance

The general rule for conventional Lagrangians is: if a Lagrangian has no explicit dependence on the coordinates then translational invariance follows accompanied by a conservation law for an energy-momentum tensor. We will show that this rule needs modification for separable Lagrangians that implement the composition of extrema.

Consider the Lagrangian:

$$\mathscr{L}_S = J\,\mathscr{L}_F(\phi(X),\, \partial\phi/\partial X^\mu) + \mathscr{L}_C(X^\mu(y),\, \partial X^\mu(y)/\partial y^\nu) \quad (2.96)$$

in which X^μ plays a dual role of both fields and coordinates. Let us consider a variation in X^μ:

$$X^\mu(y) \rightarrow X^\mu(y) + \delta X^\mu(y) \tag{2.97}$$

where $\delta X^\mu(y)$ is an arbitrary function of y that vanishes at the endpoints of the integration region of the integral. The action is:

$$I = \int \mathscr{L}_s d^4y \tag{2.98}$$

We will show that a variation in $X^\mu(y)$ leads to a conserved energy-momentum tensor. But we will use integrals of the Lagrangian density since it provides a simpler derivation of the result. Under the variation of eq. 2.97 we find

$$\delta\phi = \phi(X(y) + \delta X^\mu(y)) - \phi(X(y))$$

$$= \delta X^\mu\, \partial\phi/\partial X^\mu \tag{2.99a}$$

$$\delta(\partial\phi/\partial X^\nu) = \delta X^\mu\, \partial(\partial\phi/\partial X^\mu)/\partial X^\nu \tag{2.99b}$$

$$\delta(\partial X^\mu/\partial y^\nu) = \partial(\delta X^\mu)/\partial y^\nu \tag{2.99c}$$

The integral in eq. 2.98 changes by

$$\delta I = \int d^4y\, \delta\mathscr{L}_s = \int d^4y\, [\, \delta(J\mathscr{L}_F) + \delta\mathscr{L}_C\,] \tag{2.100a}$$

which becomes:

$$\delta I = \int d^4y\, [\delta X^\mu\, \partial(J\mathscr{L}_F)/\partial X^\mu + \partial(\delta X^\mu \partial\mathscr{L}_C/\partial(\partial X^\mu/\partial y^\nu))/\partial y^\nu] \tag{2.100b}$$

due to the equations of motion of X^μ (eq. 2.19b) in X^μs role as a set of fields. Since the second term is a total divergence its contribution to δI is zero. Thus we can express eq. 2.100b as:

$$\delta I = \int d^4y\, [J\, \delta\mathscr{L}_F + \mathscr{L}_F\, \delta J\,] \tag{2.101}$$

realizing that the Jacobian J depends on y and thus X:

$$\delta J = \delta X^\mu\, \partial J/\partial X^\mu \tag{2.102}$$

A partial integration gives

$$\mathscr{L}_F\, \delta J = \delta X^\mu\, \partial(J\mathscr{L}_F)/\partial X^\mu - \delta X^\mu\, J\, \partial\mathscr{L}_F/\partial X^\mu \tag{2.103}$$

Evaluating $\delta\mathscr{L}_F$ we find:

$$\delta\mathscr{L}_F = \partial\mathscr{L}_F/\partial\phi\, \delta\phi + \partial\mathscr{L}_F/\partial(\partial\phi/\partial X^\mu)\, \delta(\partial\phi/\partial X^\mu) \tag{2.104}$$

which gives

$$\delta\mathscr{L}_F = \delta X^\nu\, \partial/\partial X^\mu\, [\partial\mathscr{L}_F/\partial(\partial\phi/\partial X^\mu)\, \partial\phi/\partial X^\nu\,] \tag{2.105}$$

using the equations of motion eq. 2.31, and using eq. 2.99b. Combining eqs. 2.100, 2.101, 2.103 and 2.105 we obtain:

$$\int d^4y\, J\, \delta X^\nu\, \partial/\partial X_\mu\, \mathscr{T}_{F\mu\nu} = \int d^4X\, \delta X^\nu\, \partial/\partial X_\mu\, \mathscr{T}_{F\mu\nu} = 0 \tag{2.106}$$

where

$$\mathscr{T}_{F\mu\nu} = -\, g_{\mu\nu}\, \mathscr{L}_F + \partial\mathscr{L}_F/\partial(\partial\phi/\partial X_\mu)\, \partial\phi/\partial X^\nu \tag{2.107}$$

after some manipulations. Since δX^ν is an arbitrary function of y the differential conservation law follows:

$$\partial/\partial X_\mu\, \mathscr{T}_{F\mu\nu} = 0 \tag{2.108}$$

Eq. 2.108 implies the energy-momentum vector

$$P_{F\beta} = \int d^3X\, \mathscr{T}_{F0\beta} \tag{2.109}$$

is conserved:

$$\partial/\partial X^0\, P_{F\beta} = 0 \tag{2.110}$$

The hamiltonian density (eq. 2.83) is

$$\mathscr{H}_F = \mathscr{T}_{F0\beta} \tag{2.111}$$

Thus the field energy

$$H_F = P_{F0} = \int d^3X\, \mathscr{T}_{F00} \tag{2.112}$$

is conserved with respect to the "time" X^0. Later we will see that H_F is trivially conserved in the Coulomb gauge of X_μ. (We will also establish an electromagnetic-like quantum field theory for X_μ with gauge invariance.) In other gauges the conservation of H_F is not trivial.

Separable Lagrangians and Angular Momentum Conservation

We can also verify Lorentz invariance and obtain the form of the conserved angular momentum for a separable Lagrangian by considering the effect of an infinitesimal Lorentz transformation. We will consider the case of a scalar field ϕ.

Under an infinitesimal Lorentz transformation as specified by eqs. 2.60a – 2.60d the separable Lagrangian changes by

$$\delta\mathscr{L}_s = \in^{\mu\nu}\, y_\nu\, \partial\mathscr{L}_s/\partial y^\mu \tag{2.113}$$

which can also be expressed as

$$\delta\mathscr{L}_s = \partial\mathscr{L}_s/\partial\phi\; \delta\phi + \partial\mathscr{L}_s/\partial(\partial\phi/\partial X^\mu)\; \delta(\partial\phi/\partial X^\mu) + \partial\mathscr{L}_s/\partial X^\mu\; \delta X^\mu + \\ + \partial\mathscr{L}_s/\partial(\partial X^\mu/\partial y^\nu)]\; \delta(\partial X^\mu/\partial y^\nu) \tag{2.114}$$

Combining eqs. 2.113 and 2.114 leads to:

$$\in_{\mu\nu}\, \partial/\partial y^\sigma\, \mathscr{M}_s^{\ \sigma\mu\nu} = 0 \tag{2.115}$$

where

$$\mathscr{M}_s^{\ \sigma\mu\nu} = J\, \mathscr{M}_F^{\ \sigma\mu\nu} + \mathscr{M}_C^{\ \sigma\mu\nu} + \mathscr{M}_M^{\ \sigma\mu\nu} \tag{2.116}$$

$$\mathscr{M}_F^{\ \sigma\mu\nu} = (g^{\mu\sigma}y^\nu - g^{\nu\sigma}y^\mu)\mathscr{L}_F + \partial\mathscr{L}_F/\partial(\partial\phi/\partial y_\sigma)\,(y^\mu\partial\phi/\partial y_\nu - y^\nu\partial\phi/\partial y_\mu) \tag{2.117}$$

$$\mathscr{M}_C{}^{\sigma\mu\nu} = (g^{\mu\sigma}y^{\nu} - g^{\nu\sigma}y^{\mu})\mathscr{L}_C + \\ + \partial\mathscr{L}_C/\partial(\partial X^{\delta}/\partial y^{\sigma})(g^{\delta\nu}X^{\mu} - g^{\delta\mu}X^{\nu} + y^{\mu}\,\partial X^{\delta}/\partial y_{\nu} - y^{\nu}\,\partial X^{\delta}/\partial y_{\mu}) \tag{2.118}$$

$$\mathscr{M}_M{}^{\sigma\mu\nu} = \mathscr{L}_F\,\partial J/\partial(\partial X^{\delta}/\partial y^{\sigma})(g^{\delta\nu}X^{\mu} - g^{\delta\mu}X^{\nu} + y^{\mu}\,\partial X^{\delta}/\partial y_{\nu} - y^{\nu}\,\partial X^{\delta}/\partial y_{\mu}) \tag{2.119}$$

where the third term originates in the dependence of J on derivatives of X^{μ}. Eq. 2.117 was obtained in part by using the identity:

$$\partial\mathscr{L}/\partial(\partial\phi/\partial y^{\sigma}) = \partial\mathscr{L}/\partial(\partial\phi/\partial X^{\alpha})\,\partial y^{\sigma}/\partial X^{\alpha} \tag{2.120}$$

where $\mathscr{L}$ and ϕ have the form specified in eq. 2.15.

The conserved angular momentum is:

$$M_s{}^{\mu\nu} = \int dy\,\mathscr{M}_s{}^{0\mu\nu} \tag{2.121}$$

with

$$\partial M_s{}^{\mu\nu}/\partial y^0 = 0 \tag{2.122}$$

Angular Momentum and $\mathscr{L}_F$

An alternate conserved angular momentum can be obtained by considering the "field" part of the Lagrangian $\mathscr{L}_F$ under an infinitesimal Lorentz transformation ($\epsilon_{\mu\nu} = -\epsilon_{\nu\mu}$):

$$X'_{\mu} = X_{\mu} + \delta X_{\mu} \tag{2.123a}$$

$$\delta\phi = \phi(X'(y)) - \phi(X(y))$$

$$= \delta X^{\mu}\,\partial\phi/\partial X^{\mu} \tag{2.123b}$$

$$\delta X^{\mu} = S^{\mu}{}_{\alpha}X^{\alpha}(y) - X^{\mu}(y) \tag{2.123c}$$

$$= \epsilon^{\mu}{}_{\alpha}X^{\alpha}(y) \tag{2.123d}$$

where $S^{\mu}{}_{a}$ is the Lorentz transformation matrix for a vector. (If X^{μ} is a gauge field then an additional operator gauge term would have to be added to eq. 2.123d.)

The Lagrangian changes by

$$\delta \mathscr{L}_F = \epsilon^{\mu\nu} X_\nu \, \partial \mathscr{L}_F / \partial X^\mu \tag{2.124}$$

under an infinitesimal Lorentz transformation. The change can also be expressed as:

$$\delta \mathscr{L}_F = \partial \mathscr{L}_F / \partial \phi \; \delta \phi + \partial \mathscr{L}_F / \partial(\partial \phi / \partial X^\mu) \; \delta(\partial \phi / \partial X^\mu) \tag{2.125}$$

Combining eqs. 2.124 and 2.125 leads to:

$$\epsilon_{\mu\nu} \; \partial / \partial X^\sigma \; \mathscr{M}_{FX}{}^{\sigma\mu\nu} = 0 \tag{2.126}$$

where

$$\mathscr{M}_{FX}{}^{\sigma\mu\nu} = (g^{\mu\sigma} X^\nu - g^{\nu\sigma} X^\mu) \mathscr{L}_F + \partial \mathscr{L}_F / \partial(\partial \phi / \partial X^\sigma) \, (X^\mu \partial \phi / \partial X_\nu - X^\nu \partial \phi / \partial X_\mu) \tag{2.127}$$

The conserved angular momentum associated with the X coordinates is:

$$M_{FX}{}^{\mu\nu} = \int d^3X \; \mathscr{M}_{FX}{}^{0\mu\nu} \tag{2.128}$$

with

$$\partial M_{FX}{}^{\mu\nu} / \partial X^0 = 0 \tag{2.129}$$

The angular momentum density can be written in the familiar form:

$$\mathscr{M}_{FX}{}^{\sigma\mu\nu} = X^\mu \, \mathscr{T}_F^{\sigma\nu} - X^\nu \, \mathscr{T}_F^{\sigma\mu} \tag{2.130}$$

using eq. 2.107.

Separable Lagrangians and Internal Symmetries

We will now consider the case of a set of scalar fields ϕ_r in a separable Lagrangian with an internal symmetry under a local transformation

$$\phi_r(X) \rightarrow \phi_r(X) - i\epsilon\lambda_{rs}\,\phi_s(X) \qquad (2.131)$$

If the Lagrangian is invariant under this transformation, then

$$\delta\mathscr{L}_S \equiv \delta\mathscr{L}_F = 0 = \partial\mathscr{L}_F/\partial\phi_r\,\delta\phi_r + \partial\mathscr{L}_F/\partial(\partial\phi_r/\partial X^\alpha)\,\delta(\partial\phi_r/\partial X^\alpha) \qquad (2.132)$$

Using the equation of motion eq. 2.31, which is satisfied by all components ϕ_r, we obtain a conserved current:

$$\mathscr{J}^\nu = -i\,\partial\mathscr{L}_F/\partial(\partial\phi_r/\partial X^\nu)\,\lambda_{rs}\,\phi_s \qquad (2.133)$$

satisfying

$$\partial\mathscr{J}^\nu/\partial X^\nu = 0 \qquad (2.134)$$

The conserved charge is

$$Q = \int d^3X\,\mathscr{J}^0 \qquad (2.135)$$

$$\partial Q/\partial X^0 = 0\,. \qquad (2.136)$$

We note eq. 2.71 provides a corresponding conservation law for the y coordinate system.

3. Quantization of Coordinate Systems

Non-commuting Coordinates

Field theories with non-commuting coordinates are currently an active field of study.[6] Investigators are studying gauge theories, and in particular Quantum Electrodynamics, with non-commuting coordinates. Non-commuting coordinates are usually implemented quantum mechanically by positing non-zero commutators for coordinates:

$$[x^i, x^j] = i\theta^{ij} \tag{3.1}$$

The concept of non-commuting coordinates originated in the appearance of non-commuting coordinates in studies of D-brane dynamics in superstring theories.

New Approach to Non-Commuting Coordinates

In this book we will consider an alternative approach that postulates a q-number coordinate system X^μ with which all particle fields are defined. This coordinate system is realized as a mapping from a more fundamental c-number coordinate system y^ν, which we will call the subspace for want of a better term. We will treat X^μ as a vector of quantum fields, thus realizing a new type of non-commutative coordinates at unequal subspace times.

This approach is radically different from the non-commutative coordinate realizations hitherto discussed in the literature. It has a number of beneficial results to recommend it – the main result is the finiteness of quantum field theories that are defined within its framework. We will explore some of these results in the following chapters.

The X^μ coordinate system, as we define it, has a c-number real part and a q-number imaginary part. Thus particle fields which are normally defined on four-dimensional real space-time will now be defined on a complex four-dimensional space-time where four imaginary dimensions will appear as *Quantum Dimensions*™ embodied in a vector quantum field $Y^\mu(y)$.

---

[6] M. R. Douglas and N. A. Nekrasov, Rev. Mod. Phys. **73**, 977 (2002) and references therein; J. Harvey, hep-th/0102076; M. Hamanaka and K. Toda, hep-th/0211148; N. Seiberg and E. Witten, hep-th/9908142; R. J. Szabo, hep-th/0109162; G. Berrino, S. L. Cacciatori, A. Celi, L. Martucci, and A. Vicini, hep-th/0210171; S. Godfrey and A. Doncheski, DESY eprint 02-195; M. Caravati, A. Devoto, and W. W. Repko, hep-th/0211463; and references within these papers.

$$X_\mu(y) = y_\mu + i\, Y_\mu(y)/M_c^2$$

The $Y^\mu(y)$ field is a function of the subspace y coordinates. The real part of the space-time dimensions will be taken to be the subspace y coordinates.[7]

The imaginary part of space-time (which has not been experimentally seen) will simply be the quantum fluctuations of a massless vector quantum field that are suppressed further by a large mass scale – perhaps of the order of the Planck mass – that reduces the imaginary Quantum Dimensions™ to the infinitesimal. The effects of these Quantum Dimensions™ only become appreciable in quantum field theory at energies of the order of M_c. At these energies an exponential Gaussian factor in each particle (and ghost) propagator that is generated by the Quantum Dimensions™ serves to make perturbation theory calculations finite – including calculations in Quantum Gravity.

The formalism that we will describe introduces a new form of interaction that does not have the form of the simple polynomial interactions that have hitherto dominated quantum field theories. This form of interaction takes place via the composition of quantum fields and can be called a *Dimensional Interaction*™ or an *Interdimensional Interaction*™ since it couples particle behavior through Quantum Dimensions™.[8]

Quantization Using a C-Number X^μ

We will begin by considering the case of a scalar quantum field theory. We assume a real underlying y subspace. Since X^μ is a set of coordinates, we choose to define a scalar field ϕ as a function of X^μ which in turn is a function of the y^ν coordinates. We will provisionally second quantize ϕ treating X^μ as c-number coordinates using a conventional approach.[9]

We assume a Lagrangian specified by eq. 2.33 that leads to the Klein-Gordon equation, eq. 2.34. For that Lagrangian formulation, the momentum conjugate to ϕ is:

$$\pi_\phi = \partial \mathscr{L}_F / \partial \phi' \equiv \partial \mathscr{L}_F / \partial(\partial\phi/\partial X^0) \tag{3.2}$$

Following the canonical quantization procedure, π and ϕ become hermitian operators with equal time ($X^0 = X^{0\prime}$) commutation rules:

---

[7] In a deeper theory the real part might also be a quantum field that undergoes a condensation to generate c-number coordinates. We will not consider this possibility in this book.

[8] See the back of title page for the rationale for, and the details of the authorization to use, these trademarked terms.

[9] Some texts are: Bogoliubov, N. N., Shirkov, D. V., *Introduction to the Theory of Quantized Fields* (Wiley-Interscience Publishers Inc., New York, 1959); Bjorken, J. D., Drell, S. D., *Relativistic Quantum Fields* (McGraw-Hill, New York, 1965); Huang, K., *Quarks, Leptons & Gauge Fields Second Edition* (World Scientific, River Edge, NJ, 1992); Kaku, M., *Quantum Field Theory* (Oxford University Press, New York, 1993); Weinberg, S., *The Quantum Theory of Fields* (Cambridge University Press, New York, 1995).

$$[\phi(X), \phi(X')] = [\pi_\phi(X), \pi_\phi(X')] = 0 \tag{3.3}$$

$$[\pi_\phi(X), \phi(X')] = -i\,\delta^3(\mathbf{X} - \mathbf{X}') \tag{3.4}$$

The hamiltonian is defined by eq. 2.112. The standard Fourier expansion of the solution to the Klein-Gordon equation (eq. 2.34) is:

$$\phi(X) = \int d^3p\, N_m(p)\, [a(p)\, e^{-ip.X} + a^\dagger(p)\, e^{ip.X}] \tag{3.5}$$

where

$$N_m(p) = [(2\pi)^3 2\omega_p]^{-½} \tag{3.6}$$

and

$$\omega_p = (\mathbf{p}^2 + m^2)^{½} \tag{3.7}$$

The commutation relations of the Fourier coefficient operators are:

$$[a(p), a^\dagger(p')] = \delta^3(\mathbf{p} - \mathbf{p}') \tag{3.8}$$

$$[a^\dagger(p), a^\dagger(p')] = [a(p), a(p')] = 0 \tag{3.9}$$

The reader will recognize the quantization procedure is formally identical to the standard canonical quantization procedure of a free scalar quantum field.

In the case of spin ½, spin 1 and spin 2 fields the standard quantization procedure *in terms of the X coordinate system* can also be followed in a way similar to the procedure in standard texts. We will see these quantization procedures in the following chapters.

In the next section we will quantize the transformation from the y coordinate system to the X coordinate system. The quantization of the coordinates leads to a non-commutative geometry that have some superficial similarities to non-commutative geometries described in papers referenced earlier in this chapter.

Coordinate Quantization

In this section we will treat the quantization of the coordinates X^μ as a vector field defined on a fundamental c-number coordinate system y^ν of the same dimensionality. We will assume the y^ν space is a "normal" flat Minkowski space with three spatial and one time dimensions. Generalizations to spaces with more dimensions are straightforward but do not appear to be experimentally well motivated at the present time in the opinion of this author.

Thus we will assume X^μ has three spatial dimensions and one time dimension. For reasons primarily of simplicity (and to bring a string-like aspect into conventional quantum field theory), we will assume the X^μ fields are similar to the free electromagnetic vector potential A^μ with the Lagrangian:

$$\mathscr{L}_C = +\tfrac{1}{4} M_c^{\,4} F^{\mu\nu} F_{\mu\nu} \tag{3.10}$$

$$F_{\mu\nu} = \partial X_\mu/\partial y^\nu - \partial X_\nu/\partial y^\mu \tag{3.11}$$

where $M_c^{\,4}$ is a mass scale that is required on dimensional grounds and serves to set the scale for new Physics as we will see later. Note the sign in eq. 3.10 is not negative – contrary to the conventional electromagnetic Lagrangian. The reason for this difference is the solution that we wish to obtain for X^μ. Up to this point we have assumed X^μ to be real. With a view to the unified quantum field theory that we wish to construct we will posit a complex[10] solution for X^μ with the imaginary part being a quantum field $Y_\mu(y)$ and the real part being a c-number:

$$X_\mu(y) = y_\mu + i\, Y_\mu(y)/M_c^{\,2} \tag{3.12}$$

If X^μ has this form, then

$$F_{\mu\nu} = i\,(\partial Y_\mu/\partial y^\nu - \partial Y_\nu/\partial y^\mu)/M_c^{\,2} \tag{3.13}$$

Defining

$$F_{Y\mu\nu} = (\partial Y_\mu/\partial y^\nu - \partial Y_\nu/\partial y^\mu) \tag{3.14}$$

we see the Lagrangian assumes the form of the conventional electromagnetic Lagrangian:

$$\mathscr{L}_C = -\tfrac{1}{4} F_Y^{\mu\nu} F_{Y\mu\nu} \tag{3.15}$$

This Lagrangian can be used to develop field equations and a canonical quantization that is completely analogous to Quantum Electrodynamics.

---

[10] Theories of quantum mechanics, and quantum fields, in complex and quaternion spaces have been considered by numerous authors. For example see C. M. Bender, D. C. Brody and H. F. Jones, "Complex Extension of Quantum Mechanics" Phys. Rev. Letters **89**, 270401-1 (2002) and references therein; S. L. Adler and A. C. Millard, "Generalized Quantum Dynamics as Pre-Quantum Mechanics", Princeton Univ. preprint arXiv:hep-th/9508076 (1995) and references therein. These theories are all very different from the theories presented herein.

Gauge Invariance

The gauge invariance of the Lagrangian allows us to choose a convenient gauge. The gauge invariance of the full Lagrangian

$$\mathscr{L}_S = \mathscr{L}_F(\phi(X), \partial\phi/\partial X^\mu)\, J + \mathscr{L}_C(X^\mu(y), \partial X^\mu(y)/\partial y^\nu) \qquad (2.96)$$

is based on the standard gauge invariance of $\mathscr{L}_C$, and the gauge invariance of $J\mathscr{L}_F$ in the form of translational invariance

$$X^\mu(y) \rightarrow X^\mu(y) + \delta X^\mu(y) \qquad (2.97)$$

for the special case of a translation of X with the form of a gauge transformation:

$$\delta X^\mu(y) = \partial\Lambda(y)/\partial y_\mu$$

In this case eq. 2.106 implies

$$\int d^4y\, \Lambda(y)\, \partial\, [J\, \partial/\partial X^\mu\, \mathscr{T}_{F\mu\nu}]/\partial y_\nu = 0$$

after a partial integration and so gives the differential conservation law:

$$\partial\, [J\, \partial/\partial X^\mu\, \mathscr{T}_{F\mu\nu}]/\partial y_\nu = 0$$

since $\Lambda(y)$ is arbitrary. This conservation law is trivially obeyed since, by eq. 2.108:

$$\partial/\partial X^\mu\, \mathscr{T}_{F\mu\nu} = 0 \qquad (2.108)$$

Thus translational invariance in the $\mathscr{L}_F$ sector together with standard gauge invariance in the $\mathscr{L}_C$ sector automatically guarantees Y field gauge invariance of the total Lagrangian.

It appears that the most convenient gauge is the Coulomb gauge[11]:

$$\partial Y^i/\partial y^i = 0 \qquad (3.16)$$

which, in the absence of external sources, allows us to set

---

[11] It is also possible to quantize using an indefinite metric that preserves manifest Lorentz covariance as was done by Gupta and Bleuler for the electromagnetic field. We opt for manifest positivity.

$$Y^0 = 0 \tag{3.17}$$

since Y^0 does not have a canonically conjugate momentum. A conventional treatment leads to the equal time commutation relations:

$$[Y^\mu(\mathbf{y}, y^0), Y^\nu(\mathbf{y}', y^0)] = [\pi^\mu(\mathbf{y}, y^0), \pi^\nu(\mathbf{y}', y^0)] = 0 \tag{3.18}$$

$$[\pi^j(\mathbf{y}, y^0), Y_k(\mathbf{y}', y^0)] = -i\,\delta^{tr}{}_{jk}(\mathbf{y} - \mathbf{y}') \tag{3.19}$$

(note the locations of the j indexes in eq. 3.19 introduce a minus sign) where

$$\pi^k = \partial \mathscr{L}_C / \partial Y_k' \tag{3.20}$$

$$\pi^0 = 0 \tag{3.21}$$

$$\delta^{tr}{}_{jk}(\mathbf{y} - \mathbf{y}') = \int d^3k\, e^{i\,\mathbf{k}\bullet(\mathbf{y} - \mathbf{y}')} (\delta_{jk} - k_j k_k/\mathbf{k}^2)/(2\pi)^3 \tag{3.22}$$

$$Y_k' = \partial Y_k/\partial y^0 \tag{3.23}$$

The Coulomb gauge reveals the two degrees of freedom that are present in the vector potential. The Fourier expansion of the vector potential is:

$$Y^i(y) = \int d^3k\, N_0(k) \sum_{\lambda=1}^{2} \varepsilon^i(k, \lambda)[a(k,\lambda)\, e^{-ik.y} + a^\dagger(k,\lambda)\, e^{ik.y}] \tag{3.24}$$

where

$$N_0(k) = [(2\pi)^3 2\omega_k]^{-1/2} \tag{3.25}$$

and (m = 0)

$$\omega_k = (\mathbf{k}^2)^{1/2} = k^0 \tag{3.26}$$

with $\vec{\epsilon}(k, \lambda)$ being the polarization unit vectors for $\lambda = 1,2$ and $k^\mu k_\mu = 0$.

The commutation relations of the Fourier coefficient operators are:

$$[a(k,\lambda), a^\dagger(k',\lambda')] = \delta_{\lambda\lambda'}\,\delta^3(\mathbf{k} - \mathbf{k}') \tag{3.27}$$

$$[a^\dagger(k,\lambda),\ a^\dagger(k',\lambda')] = [a(k,\lambda),\ a(k',\lambda')] = 0 \qquad (3.28)$$

and the polarization vectors satisfy

$$\sum_{\lambda=1}^{2} \varepsilon_i(k,\ \lambda)\varepsilon_j(k,\ \lambda) = (\delta_{ij} - k_i k_j/\mathbf{k}^2) \qquad (3.29)$$

It will be convenient to divide the Y field into positive and negative frequency parts:

$$Y^{+}{}_i(y) = \int d^3k\ N_0(k) \sum_{\lambda=1}^{2} \varepsilon_i(k,\ \lambda)\ a(k,\lambda)\ e^{-ik.y} \qquad (3.30)$$

and

$$Y^{-}{}_i(y) = \int d^3k\ N_0(k) \sum_{\lambda=1}^{2} \varepsilon_i(k,\ \lambda)\ a^\dagger(k,\lambda)\ e^{ik.y} \qquad (3.31)$$

For later use we note the commutator between the positive and negative frequency parts is:

$$[\ Y^{-}{}_j(y_1),\ Y^{+}{}_k(y_2)] = -\int d^3k\ e^{ik\bullet(y_1 - y_2)}\ (\delta_{jk} - k_j k_k/\mathbf{k}^2)/[(2\pi)^3 2\omega_k] \qquad (3.32)$$

ϕ Particle States

We now turn to the ϕ particle states. The raising and lowering operators can be used to define free particle states. For example a one-particle state of momentum p is

$$|p\rangle = a^\dagger(p)|0\rangle \qquad (3.33)$$

with corresponding bra state

$$\langle p| = \langle 0|a(p) \qquad (3.34)$$

where the vacuum is defined as usual:

$$a(p)|0\rangle = 0 \qquad (3.35)$$

$$\langle 0|a^\dagger(p) = 0 \qquad (3.36)$$

Multi-particle states can also be defined in the conventional way with products of raising and lowering operators applied to the vacuum.

Y Fock Space Imaginary Coordinate States

States can also be defines for the quantized Y field. These states will be similar in form to electromagnetic photon states but play a different role in our approach since they are in fact coordinate excitation states for the imaginary part of X^μ. Thus the scalar field (and other particle fields) will exist in a real four-dimensional space with quantum excitations into imaginary Quantum Dimensions™. These excitations become significant at high energies. At the low energies with which we are familiar, space-time appears real; at very high energies space-time becomes slightly complex.

There are two types of imaginary coordinate excitations: 1.) Quantum excitations into Fock states consisting of superpositions of states with a definite finite number of Y "particles" and 2.) Imaginary coordinate excitations into coherent Y states with an "infinite" number of particles. Coherent states can be viewed as representing "classical" fields.

In this section we will consider Y field states with a definite number of excitations ("particles"). The raising and lowering operators of the Y field can be used to define free particle states. For example a one particle state can be defined by

$$|k, \lambda> = a^\dagger(k, \lambda)|0> \tag{3.37}$$

with corresponding bra state

$$<k, \lambda| = <0|a(k, \lambda) \tag{3.38}$$

where the "coordinate vacuum" is defined as usual:

$$a(k, \lambda)|0> = 0 \tag{3.39}$$

$$<0|a^\dagger(k, \lambda) = 0 \tag{3.40}$$

Multi-particle states can also be defined in the conventional way with products of the raising and lowering operators applied to the vacuum. The set of all states containing a finite number of "particles" constitutes a Fock space.

A state with a finite number of Y "particles" represents a quantum fluctuation into imaginary Quantum Dimensions™.

Y Coherent Imaginary Coordinate States

Coherent Y states bring us closer what we might consider to be "classical" imaginary dimensions – dimensions that we can, in principle, experience as we do normal dimensions. Let us define the coherent state[12]

$$| y, p> = e^{-p.Y^-(y)/M_c^2}|0> \tag{3.41}$$

This state is an eigenstate of the coordinate operator $Y^+(y')$:

$$Y^+_j(y_1)\ |y_2, p> = -[Y^+_j(y_1), \mathbf{p}\cdot\mathbf{Y}^-(y_2)]/M_c^2\ |y, p> \tag{3.42}$$

$$= -\int d^3k\ [N_0(k)]^2\ e^{ik\bullet(y_2 - y_1)}\ (p_j - k_j\mathbf{p}\cdot\mathbf{k}/\mathbf{k}^2)/M_c^2\ |y, p> \tag{3.43}$$

$$= p^i\Delta_{Tij}(y_1 - y_2)/M_c^2\ |y, p> \tag{3.44}$$

where $p^i\Delta_{Tij}(y_1 - y_2)/M_c^2$ is the eigenvalue of $Y^+_j(y_1)$. As we will see in the next chapter, the eigenvalue of Y^+ becomes large as $(y_1 - y_2)^2 \to 0$. Thus the imaginary Quantum Dimensions™ become significant at very short distances, and significantly modify the high-energy behavior of quantum field theories. In particular, Quantum Dimensions™ have a significant effect when

$$(y_1 - y_2)^2 \lessapprox (4\pi^2 M_c^2)^{-1} \tag{3.45}$$

according to eq. 4.13 in the next chapter. We are assuming the mass scale M_c is very large – perhaps of the order of the Planck mass (1.221×10^{19} GeV/c^2). Thus imaginary Quantum Dimensions™ are far from being detectable in today's "low" energy experiments. Their effect could be of significance in the analysis of the first instants after the Big Bang.

The cover of this book has an artist's depiction of imaginary Quantum Dimensions™ "fluffing out" the real dimensions of space.

The Dynamical Generation of New Dimensions

Effectively, the imaginary dimensions that we have constructed raise the total number of real and Quantum Dimensions™ to 8 with 6 space dimensions and two time dimensions. As we will see later the requirement of gauge invariance for the quantized Y field reduces the number of time dimensions to one and constrains the six space dimensions to five degrees of

[12] Coherent states are well known in the physics literature. See for example T. W. B. Kibble, J. Math. Phys. **9**, 315 (1968) and references therein; V. Chung, Phys. Rev. **140**, B1110 (1965); J. R. Klauder, J. McKenna, and E. J. Woods, J. Math. Phys. **7**, 822 (1966) and references therein.

freedom. Since X is a function of y we can also view the four dimensional world that we live in as a four-dimensional surface in an eight-dimensional space-time.

Generation of Quantum Dimensions™ by the ϕ (X) field

The ϕ(X) field generates Quantum Dimensions™ via coherent states from the vacuum. From eq. 3.5 and 3.12 we see

$$\phi(X) = \int d^3p\ N_m(p)\ [a(p)\ e^{-ip\cdot(y + iY/M_c^2)} + a^\dagger(p)\ e^{ip\cdot(y + iY/M_c^2)}] \qquad (3.46)$$

with the result

$$\phi(X)|0\rangle = \int d^3p\ N_m(p)\ a^\dagger(p)\ e^{ip\cdot(y + iY/M_c^2)}|0\rangle \qquad (3.47)$$

is a superposition of coherent Y states plus one scalar particle.

We can also define coherent Y states with total momentum q using the expression:

$$|q\ Y\rangle = \int d^4y\ e^{iq\cdot X(y)}|0\rangle = \int d^4y\ e^{iq\cdot(y + iY/M_c^2)}|0\rangle \qquad (3.48)$$

Expanding the Y part of the exponential in eq. 3.48 gives

$$|q\ Y\rangle = \sum_{n=0}^{\infty}(-1)^n(n!)^{-1}\prod_{j=1}^{n}\left(\int d^3k_j N_0(k_j)\right)\delta^4(q - \sum_{s=1}^{n} k_s)\prod_{r=1}^{n}\sum_{\lambda_r=1}^{2} \mathbf{q}\cdot \varepsilon(k_r, \lambda_r)\ a^\dagger(k_r,\lambda_r)|0\rangle$$

$$(3.49)$$

which shows that the sum of the Y particle momenta for each term in the expansion is q.

Hamiltonian for Particle and Coordinate States

The hamiltonian for the separable, coordinate quantized, scalar quantum field theory is:

$$H_s = \int \mathscr{H}_s\ d^3y \qquad (2.79)$$

with

$$\mathscr{H}_s = J\ \mathscr{H}_F + \mathscr{H}_C \qquad (2.82)$$

$$\mathscr{H}_F(\ \phi(X),\ \pi_\phi,\ \partial\phi/\partial X^i) = \pi_\phi\ \phi' - \mathscr{L}_F \qquad (2.83)$$

$$\mathscr{H}_C(X^\mu(y), \pi_X{}^\mu, \partial X^\mu(y)/\partial y^j, y^\nu) = \pi_X{}^\mu X_\mu{}' - \mathscr{L}_C \tag{2.84}$$

$$\mathscr{L}_F = \tfrac{1}{2}[(\partial\phi/\partial X^\nu)^2 - m^2\phi^2] \tag{2.33}$$

$$\mathscr{L}_C = +\tfrac{1}{4} M_c{}^4 F^{\mu\nu}F_{\mu\nu} \tag{3.10}$$

We note

$$\mathscr{H}_F = \tfrac{1}{2}[\pi_\phi{}^2 + (\partial\phi/\partial X^i)^2 + m^2\phi^2] \tag{3.50}$$

is the conventional scalar particle hamiltonian when viewed as a function of the X coordinates. $\mathscr{H}_C$ has the same form as the conventional electromagnetic hamiltonian when eq. 3.12 is used to specify X in terms of the Y fields.

$$\mathscr{H}_C = \tfrac{1}{2}(E_Y{}^2 + B_Y{}^2) \tag{3.51}$$

where

$$E_Y{}^i = -\partial Y^i/\partial y^0 \tag{3.52}$$

$$B_Y{}^i = \varepsilon^{ijk}\, \partial Y_j/\partial y^k \tag{3.53}$$

Using the fourier expansions of ϕ and X^μ (eqs. 3.5 and 3.24) we obtain the following expression for the normal-ordered hamiltonian H_S:

$$P_s{}^0 \equiv H_s = \int :\mathscr{H}_s: d^3y \tag{3.54}$$

$$H_s = \int d^3p\,(\mathbf{p}^2 + m^2)^{1/2} a^\dagger(p)a(p) + \int d^3k \sum_{\lambda=1}^{2} (\mathbf{k}^2)^{1/2} a^\dagger(k,\lambda)a(k,\lambda) \tag{3.55}$$

where : : indicates normal ordering and where we perform a functional integration over X (Note the Jacobian is present.) for the particle part of the hamiltonian $\mathscr{H}_F$. The hamiltonian is manifestly positive definite.

The spatial momentum is specified by

$$P_s^j \;= -\int d^3X :\pi_\phi(X)\partial\phi(X)/\partial X_j: + \int d^3y :E_Y{}^i\partial Y^i/\partial y_j: \qquad (3.56)$$

$$= \int d^3p\; p^j\, a^\dagger(p)a(p) + \int d^3k \sum_{\lambda=1}^{2} k^j\, a^\dagger(k, \lambda)a(k, \lambda) \qquad (3.57)$$

The momentum operator generates displacements in ϕ

$$[P_s{}^\mu, \phi(X)] = -\, i\partial\phi/\partial X_\mu \qquad (3.58)$$

Second Quantized Coordinates

At this point we have developed a formalism for a scalar particle quantum field theory based on non-commutative coordinates. In the following chapters we will proceed to use this formalism to develop a unified quantum field theory of the known forces of nature.

4. Scalar Two-Tier Quantum Field Theory

It appears then that there must be fundamental changes in our basic formulation of quantum field theory, so that unrenormalized masses and unrenormalized coupling constants can become finite.
T. D. Lee & G. C. Wick[13]

Introduction

There are a number of reasons to believe that particles have a string-like aspect under certain experimental conditions. This string-like aspect of their nature apparently manifests itself at extremely short distances.

With these thoughts in mind we will examine a new formulation of quantum field theory that we will call *two-tier quantum field theory*. This type of quantum field theory incorporates something like a string-like substructure within a standard quantum field theoretic framework. In this chapter we will apply our approach to scalar particle quantum field theory. In the following chapters we will apply the approach to QED, Electroweak Theory, the Standard Model and lastly a unified model for the known forces of nature.

"Two-Tier" Space

In the preceding chapter we developed a quantized coordinate system X^μ defined on an underlying c-number coordinate system y^ν:

$$X_\mu(y) = y_\mu + iY_\mu(y)/M_c^2 \qquad (3.12)$$

$$Y^i(y) = \int d^3k\, N_0(k) \sum_{\lambda=1}^{2} \varepsilon^i(k, \lambda)[a(k,\lambda)\, e^{-ik\cdot y} + a^\dagger(k,\lambda)\, e^{ik\cdot y}] \qquad (3.24)$$

We also developed a conventional scalar quantum field theory with the Fourier expansion:

---

[13] T. D. Lee and G. C. Wick, Phys. Rev. **D2**, 1033 (1970). Lee and Wick's model QED is totally unrelated to our approach.

$$\phi(X) = \int d^3p\, N_m(p)\, [a(p)\, e^{-ip.X} + a^\dagger(p)\, e^{ip.X}] \qquad (3.5)$$

We will now consider the implications of the separable Lagrangian:

$$\mathscr{L}_S = \mathscr{L}_F(\phi(X), \partial\phi/\partial X^\mu)\, J + \mathscr{L}_C(X^\mu(y), \partial X^\mu(y)/\partial y^\nu) \qquad (2.96)$$

where

$$\mathscr{L}_F = ½\, [\, (\partial\phi/\partial X^\nu)^2 - m^2\phi^2\,] \qquad (2.33)$$

and

$$\mathscr{L}_C = +¼\, M_c^{\,4} F^{\mu\nu} F_{\mu\nu} \qquad (3.10)$$

or alternately,

$$\mathscr{L}_C = -¼\, F_Y^{\ \mu\nu} F_{Y\mu\nu} \qquad (3.15)$$

with

$$F_{Y\mu\nu} = \partial Y_\mu/\partial y^\nu - \partial Y_\nu/\partial y^\mu \qquad (3.14)$$

M_c is the mass that sets the scale at which the imaginary part of X^μ becomes significant.

We can view this quantum field theory as a conventional quantum field theory until energies reach the magnitude of M_c. At energies of the order of M_c, and above, the imaginary part of X^μ becomes significant and alters the high-energy behavior of the theory in a major way. This modification eliminates divergences that normally appear in perturbation theory when interactions are introduced. Yet the low energy behavior of the theory remains the same. Thus the precise calculations of QED that have been verified to an amazing degree of accuracy remain valid when a two-tier formulation of QED is created (in chapter 5). And the "low energy" results found in other conventional quantum field theories such as Electroweak Theory and the Standard Model also are closely approximated by their corresponding two-tier versions.

The straightforward use of the above equations[14] (and the canonical quantization described in the preceding chapters) leads to a scalar quantum field with the Fourier expansion:

$$\phi(X) = \int d^3p\, N_m(p)\, [a(p)\, e^{-ip.(y + iY/M_c^{\,2})} + a^\dagger(p)\, e^{ip.(y + iY/M_c^{\,2})}] \qquad (4.1)$$

[14] The use of functionals in quantum field theory is, of course, far from new as one can see in texts such as Bogoliubov and Shirkov (for example pp. 198-226).

using eq. 3.5 above. We note the equal time commutation relations of ϕ and π_ϕ are the same as the conventional equal time commutation relations of a scalar field despite the fact that X^μ and Y^μ are themselves quantum fields since $[Y^\mu(\mathbf{y}, y^0), Y^\nu(\mathbf{y}', y^0)] = 0$. In addition, the ϕ and π_ϕ fields are now not hermitean.

Armed with this information we can follow the same heuristic procedure that we followed in the preceding chapter to obtain the hamiltonian of eq. 3.50.

The Fourier expansion of ϕ does require one refinement – the exponential terms in X^μ must be normal ordered to avoid infinities in the unequal time commutation relations:

$$\phi(X) = \int d^3p\, N_m(p)\, [a(p)\, :e^{-ip\cdot(y + iY/M_c^2)}: + a^\dagger(p)\, :e^{ip\cdot(y + iY/M_c^2)}:] \tag{4.2}$$

Since the hamiltonian as well as other quantities are normal ordered in quantum field theory the additional requirement of normal ordering in the field operator is merely an extension of the standard procedure to a more complex situation and is not disturbing. The unequal time commutation relation of the normal ordered ϕ field is:

$$[\phi(X^\mu(y_1)), \phi(X^\mu(y_2))] = i\Delta(y_1 - y_2) + \mathcal{O}(1/M_c^2) \tag{4.3}$$

where

$$\Delta(y_1 - y_2) = -i \int d^3k\, (e^{-ik\bullet(y_1 - y_2)} - e^{ik\bullet(y_1 - y_2)})/[(2\pi)^3 2\omega_k] \tag{4.4}$$

is a familiar c-number invariant singular function. The additional terms in eq. 4.3 are q-number terms that become significant at very short distances of the order M_c^{-1}. Thus precise measurements of field strengths at larger distances are limited by standard quantum effects as indicated by the commutation relation.

The principle of *microscopic causality* is violated at extremely short distances of the order M_c^{-1} since the commutator (eq. 4.3) is non-zero, in general, for space-like distances of the order of M_c^{-1} due to the q-number terms. This violation is not measurable now – and for the foreseeable future – and reflects a type of non-locality at extremely short distances.

The short distance behavior of two-tier quantum field theory leads to the elimination of divergences and finite interacting quantum field theories.

Vacuum Fluctuations

While the expectation value of a *conventional* free scalar field $\phi_{conv}(X)$ is zero in a conventional quantum field theory:

$$<0|\phi_{conv}(X)|0> = 0 \tag{4.5}$$

the vacuum fluctuations of *conventional* scalar quantum field theory are quadratically divergent:

$$<0|\phi_{conv}(X)\phi_{conv}(X)|0> = \int d^3p/[(2\pi)^3 2\omega_p] \tag{4.6}$$

In "two-tier" quantum field theory we find the vacuum expectation value of a free field is zero (like eq. 4.5) *and the expectation value of the square of the field is also zero:*

$$<0|\phi(X)\phi(X)|0> = \int d^3p\, e^{-p^ip^j\Delta_{Tij}(0)}/[(2\pi)^3 2\omega_p] = 0 \tag{4.7}$$

since the exponential factor in the integral is $-\infty$. The exponent contains

$$\Delta_{Tij}(z) = \int d^3k\, e^{-ik\bullet z}\, (\delta_{ij} - k_ik_j/\mathbf{k}^2)/[(2\pi)^3 2\omega_k] \tag{4.8}$$

where "T"is for "Two-Tier". Thus *vacuum fluctuations are zero in two-tier quantum field theory.* Correspondingly, we will see that renormalization constants are finite in the two-tier versions of QED, Electroweak Theory, the Standard Model and Quantum Gravity.

The Feynman Propagator

Calculations in the perturbation theory of an interacting quantum field theory use Feynman propagators. The Feynman propagator for a two-tier free scalar quantum field is:

$$i\Delta_F^{TT}(y_1 - y_2) = <0|T(\phi(X(y_1)),\phi(X(y_2)))|0> \tag{4.9}$$

$$\equiv <0|\phi(X(y_1))\phi(X(y_2))|0>\, \theta(y_1^0 - y_2^0) +$$

$$+\, \phi(X(y_2))\phi(X(y_1))|0>\, \theta(y_2^0 - y_1^0) \tag{4.10}$$

Since $X^0 = y^0$ in the Coulomb gauge of the X^μ field there is no ambiguity in the choice of the relevant time variable. A straightforward calculation shows:

$$i\Delta_F^{TT}(y_1 - y_2) = i \int d^4p\, e^{-ip\bullet(y_1 - y_2)}\, R(\mathbf{p}, y_1 - y_2)/[(2\pi)^4(p^2 - m^2 + i\varepsilon)] \tag{4.11}$$

where

$$R(\mathbf{p}, y_1 - y_2) = \exp[-p^ip^j\Delta_{Tij}(y_1 - y_2)/M_c^4] \tag{4.12}$$

$$= \exp\{-p^2[A(v) + B(v)\cos^2\theta] / [4\pi^2 M_c^4 z^2]\} \quad (4.13)$$

with

$$z^\mu = y_1^\mu - y_2^\mu \quad (4.14)$$

$$z = |\mathbf{z}| = |\mathbf{y_1} - \mathbf{y_2}| \quad (4.15)$$

$$p = |\mathbf{p}| \quad (4.16)$$

$$v = |z^0|/z \quad (4.17)$$

$$A(v) = (1 - v^2)^{-1} + .5v \ln[(v - 1)/(v + 1)] \quad (4.18)$$

$$B(v) = v^2(1 - v^2)^{-1} - 1.5v \ln[(v - 1)/(v + 1)] \quad (4.19)$$

$$\mathbf{p} \cdot \mathbf{z} = pz \cos\theta \quad (4.20)$$

and $|\mathbf{p}|$ denoting the length of a spatial vector **p** while $|z^0|$ is the absolute value of z^0.

As eq. 4.11 indicates, the Gaussian damping factor R(p, z) for large momentum p is the same for both the positive and negative frequency parts of the two-tier Feynman propagator. It is also important to note that R(p, z) does not depend on p^0 (in the Y Coulomb gauge) and thus the integration over p^0 proceeds in the usual way to produce time-ordered positive and negative frequency parts.

Large Distance Behavior of Two-Tier Theories

The large distance behavior of the two-tier Feynman propagator approaches the behavior of the conventional Feynman propagator since

$$R(\mathbf{p}, y_1 - y_2) \to 1 \quad (4.21)$$

when $(y_1 - y_2)^2$ becomes much larger than M_c^{-2} as eq. 4.11 shows. Thus the behavior of a conventional quantum field theory naturally emerges at large distance. We will see that the conventional Standard Model is the large distance limit of the two-tier Standard Model thus *realizing a form of Correspondence Principle for Quantum Field Theory*. Features of the conventional Standard Model that depend specifically on the existence of divergences, such as the axial anomaly, will be different in the two-tier Standard Model since it is a finite theory.

Short Distance Behavior of Two-Tier Theories

At short distances the Gaussian factor dominates and radically changes the behavior of the Feynman propagator eliminating its short distance singular behavior, and thus paving the way to finite quantum field theories. Near the light cone $M_c^{-2} \gg -(y_1 - y_2)^2 \rightarrow 0$ we can approximate eq. 4.11 with

$$i\Delta_F^{TT}(y_1 - y_2) \approx \int d^3p\, [N(p)]^2\, R(\mathbf{p}, y_1 - y_2) \qquad (4.22)$$

since $e^{-ip\bullet(y_1 - y_2)}$ is approximately unity for small $(y_1 - y_2)$. We assume the mass of the ϕ particle is zero or negligible at high energies. Upon performing the integrations for space-like $(y_1 - y_2)^2$ (and analytically continuing to the time-like regions[15,16]) in eq. 4.22 we find

$$i\Delta_F^{TT}(y_1 - y_2) \approx [z^2 M_c^4/(4i\sqrt{A}\sqrt{B})] \ln[(\sqrt{A} + i\sqrt{B})/(\sqrt{A} - i\sqrt{B})] \qquad (4.23)$$

with A and B defined in eqs. 4.18 and 4.19. As $(y_1 - y_2)^2 \rightarrow 0$ from the space-like or time-like side of the light cone we find eq. 4.23 becomes:

$$i\Delta_F^{TT}(y_1 - y_2) \rightarrow \pi M_c^4 |(y_1 - y_2)^\mu (y_1 - y_2)_\mu|/8 \qquad (4.24)$$

Eq. 4.24 has several noteworthy points:

1. The propagator is well behaved on the light cone and approaches zero smoothly from both space-like and time-like directions. In contrast, the conventional scalar Feynman propagator diverges as $[(y_1 - y_2)^\mu (y_1 - y_2)_\mu]^{-2}$. This good behavior near the light cone will be seen later for other particle propagators with the net result that the usual infinities found in conventional quantum field theory are absent in two-tier quantum field theories.

2. The quadratic form of the propagator in eq. 4.24 is suggestive of attempts to formulate a relativistic harmonic oscillator model of elementary particles[17] and more recent attempts to achieve quark confinement. The fact that the absolute value of the quadratic term appears in eq. 4.24 neatly avoids the common pitfall seen in fully relativistic harmonic oscillator attempts.

---

[15] See S. Blaha, "Relativistic Bound State Models with Quasi-Free Constituent Motion", Phys. Rev. **D12**, 3921 (1975) and references therein.

[16] It should be noted that A and B in eq. 4.23 have the same sign for $0 \leq v < 1.1243$ thus making for easy analytic continuation across the light cone (which corresponds to v = 1 in eqs. 4.18 and 4.19).

[17] H. Yukawa, H., Phys. Rev. **91**, 416 (1953); Y. S. Kim and M. E. Noz, Phys. Rev. **D8**, 3521 (1973) and references therein.

3. The quadratic behavior *in coordinate space* of the propagator at short distances is equivalent to a high-energy behavior of

$$p^{-6} \tag{4.25}$$

in momentum space. Thus we get the *benefits of a higher derivative theory* in two-tier quantum field theory while retaining a positive definite energy spectrum. The problems of negative metric states that have plagued conventional higher derivative quantum field theories are avoided.[18]

String-like Substructure of the Theory

The imaginary Quantum Dimensions™ endow a particle with an extended structure that resembles to some extent the extended structure seen in bosonic string and Superstring theories. For example, Bailin And Love use the operator[19]

$$V_\Lambda(k) = \int d^2\sigma \ \sqrt{-h} \ W_\Lambda(\tau, \sigma)\, e^{-ik.X} \tag{4.26}$$

where X^μ is a quantized fourier expansion of the string fields (see eq. 7.22 of Bailin and Love).

We note our X^μ coordinate-field has two transverse degrees of freedom due to gauge invariance, which also invites comparison to the bosonic string. A point of difference is that we will create a well-defined quantum field theoretic formulation in conventional space-time that has the Standard Model as its "large distance" behavior thus introducing a note of reality that is not (yet?) apparent in Superstring theories. We see that the interacting quantum field theories based on this approach also have good, finite, short distance behavior just as string theories.

The scalar, and other particles', Feynman propagators can be viewed as describing the propagation of a particle cloaked (accompanied) by a cloud of Y particles (which generates the $R(\mathbf{p}, y_1 - y_2)$ factor in the propagator of eq. 4.11). If we examine the fourier transform of R(p, z) we see:

$$(2\pi)^4 R(\mathbf{p}, q) = \int d^4z\, e^{iq\bullet z} R(p, \mathbf{z}) = \int d^4z\, e^{iq\bullet z} \exp[-p^ip^j\Delta_{Tij}(z)/M_c^4] \tag{4.27}$$

---

[18] S. Blaha, Phys.Rev. **D10**, 4268 (1974); S. Blaha, Phys.Rev. **D11**, 2921 (1975); S. Blaha, Nuovo Cim. **A49**, :113 (1979); S. Blaha, "Generalization of Weyl's Unified Theory to Encompass a Non-Abelian Internal Symmetry Group" SLAC-PUB-1799, Aug 1976; S. Blaha, "Quantum Gravity and Quark Confinement" Lett. Nuovo Cim. **18**, 60 (1977); Nakanishi, N., Suppl. Prog. Theo. Phys. **51**, 1 (1972); and references therein.

[19] D. Bailin and A. Love, *Supersymmetric Gauge Field Theory and String Theory* (Institute of Physics Publishing, Philadelphia, PA, 1994) page 272.

we find

$$R(\mathbf{p},q) = \sum_{n=0}^{\infty} [i(2\pi M_c)^4]^{-n} (n!)^{-1} \prod_{j=1}^{n} [\int d^4k_j\, \theta(k_j^0)(\mathbf{p}^2 - (\mathbf{p}\cdot\mathbf{k}_j)^2/\mathbf{k}_j^2)/(k_j^2 + i\varepsilon)]\, \delta^4(q - \sum_r k_r) \tag{4.28}$$

which can be interpreted as a "cloud" of Y particles dressing the "bare" particle propagator. (The manifest divergences in eq. 4.28 for R(p, q) are an artifact of the expansion and the subsequent fourier transformation. They are not present in the $R(\mathbf{p}, y_1 - y_2)$ factor in the propagator of eq. 4.11.) See Fig. 4.1 for the Feynman diagram of the two-tier cloaked propagator as compared to the normal scalar particle Feynman propagator.

The two-tier Feynman propagator is basically a conventional scalar propagator that is modified by coherent Y particle emission.[20]

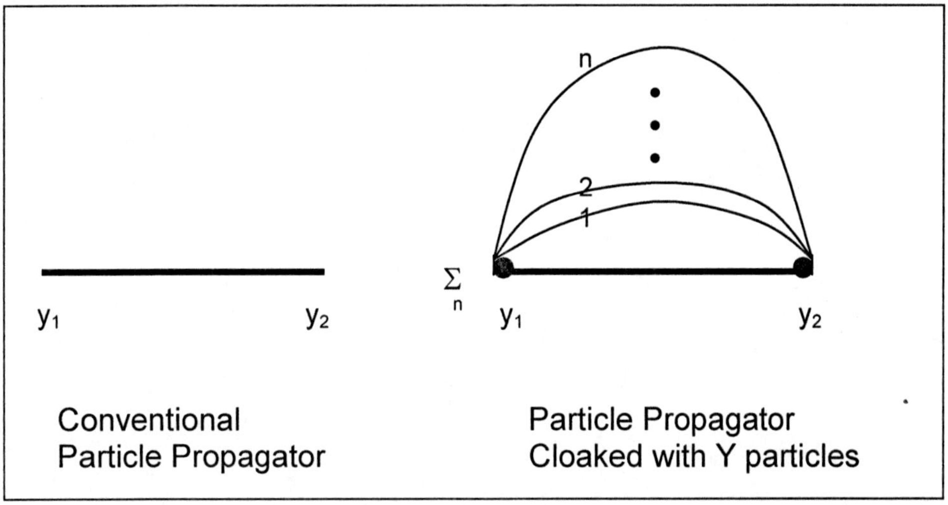

Figure 4.1. Feynman diagram for conventional and cloaked two-tier propagators.

We note that R(p, q) satisfies the convolution theorem:

$$\int d^4k\, R(\mathbf{p}, k)\, R(\mathbf{p}, q-k) = [R(\mathbf{p}, q)]^2 \tag{4.29a}$$

or

[20] T. W. B. Kibble, Phys. Rev. **173**, 1527 (1968) and references therein. In particular see page 1532 of Kibble's paper.

$$(2\pi)^4 \int d^4z\, e^{iq\bullet z} R(p, \mathbf{z}) R(p, \mathbf{z}) = [\int d^4z\, e^{iq\bullet z} R(\mathbf{p}, z)]^2 \quad (4.29b)$$

The proof follows from eq. 4.28 and the Binomial theorem.

Parity

Parity can appear in two guises within the framework of two-tier quantum field theory. One can consider a parity operation where the space parts of X^μ are reversed while y^μ is unchanged. Or one can consider a second type of parity where the space parts of y^μ are reversed.

X Parity

Under this form of parity operation y^μ is unchanged while the arguments of ϕ *appear* to change by

$$X^i(y) \rightarrow -X^i(y) \quad (4.30)$$

$$X^0(y) \rightarrow X^0(y) \quad (4.31)$$

We will denote the parity operator of this type $\mathscr{P}_X$. Under $\mathscr{P}_X$ the arguments of the scalar quantum field operator ϕ change according to eqs. 4.30-1 so that ϕ transforms as

$$\mathscr{P}_X\phi(\mathbf{X}(y), X^0(y))\mathscr{P}_X^{-1} = \phi(-\mathbf{X}(y), X^0(y)) \quad (4.32)$$

From the form of ϕ in eq. (4.2) we see we can implement eq. 4.32 by requiring:

$$\mathscr{P}_X a(\mathbf{p}, p^0)\mathscr{P}_X^{-1} = a(-\mathbf{p}, p^0) \quad (4.33)$$

$$\mathscr{P}_X a^\dagger(\mathbf{p}, p^0)\mathscr{P}_X^{-1} = a^\dagger(-\mathbf{p}, p^0) \quad (4.34)$$

$$\mathscr{P}_X X^0(y)\mathscr{P}_X^{-1} = X^0(y) \quad (4.35)$$

$$\mathscr{P}_X X^i(y)\mathscr{P}_X^{-1} = X^i(y) \quad (4.36)$$

$$\mathscr{P}_X Y^i(y)\mathscr{P}_X^{-1} = Y^i(y) \quad (4.37)$$

where i = 1,2,3.

This parity transformation is analogous to the standard form of parity transformation in conventional quantum field theory. The separable Lagrangian in eq. 2.96 (and listed at the beginning of this chapter) is invariant under this parity transformation.

y Parity

This form of parity transformation in which $y^i \to -y^i$ has significant differences from the normal parity transformation. We specify this parity transformation for a scalar quantum field by:

$$\mathscr{P}_y\phi(\mathbf{X}(\mathbf{y}, y^0), X^0(\mathbf{y}, y^0))\mathscr{P}_y^{-1} = \phi(\mathbf{X}(-\mathbf{y}, y^0), X^0(-\mathbf{y}, y^0)) \qquad (4.38)$$

This transformation can be implemented through the following set of transformations:

$$\mathscr{P}_y a(\mathbf{p}, p^0)\mathscr{P}_y^{-1} = a(-\mathbf{p}, p^0) \qquad (4.39)$$

$$\mathscr{P}_y a^\dagger(\mathbf{p}, p^0)\mathscr{P}_y^{-1} = a^\dagger(-\mathbf{p}, p^0) \qquad (4.40)$$

$$\mathscr{P}_y X^0(\mathbf{y}, y^0)\mathscr{P}_y^{-1} = X^0(-\mathbf{y}, y^0) \qquad (4.41)$$

$$\mathscr{P}_y Y^i(\mathbf{y}, y^0)\mathscr{P}_y^{-1} = -Y^i(-\mathbf{y}, y^0) \qquad (4.42a)$$

$$\mathscr{P}_y a(\mathbf{k}, k^0, 1)\mathscr{P}_y^{-1} = a(-\mathbf{k}, k^0, 1) \qquad (4.42b)$$

$$\mathscr{P}_y a(\mathbf{k}, k^0, 2)\mathscr{P}_y^{-1} = -a(-\mathbf{k}, k^0, 2) \qquad (4.42c)$$

where i = 1,2,3 and assuming the convention: $\varepsilon(\mathbf{k},k^0,1)=-\varepsilon(-\mathbf{k},k^0,1)$ and $\varepsilon(\mathbf{k},k^0,2)=+\varepsilon(-\mathbf{k},k^0,2)$.

Forms of the Parity Transformations

The parity transformations for a scalar particle are

$$\mathscr{P}_X = \exp\{-i\pi\int d^3p\, [a^\dagger(\mathbf{p}, p^0)a(\mathbf{p}, p^0) - a^\dagger(\mathbf{p}, p^0)a(-\mathbf{p}, p^0)]/2\} \qquad (4.43a)$$

$$\mathscr{P}_y = \mathscr{P}_X \exp\{-i\pi \int d^3k\, [\sum_{\lambda=1}^{2} a^\dagger(\mathbf{k},k^0,\lambda)a(\mathbf{k},k^0,\lambda) - a^\dagger(\mathbf{k}, k^0, 1)a(-\mathbf{k},k^0,1) +$$

$$+ a^\dagger(\mathbf{k}, k^0, 2)a(-\mathbf{k},k^0,2)]/2\} \qquad (4.43b)$$

The separable Lagrangian of eq. 2.96 is invariant under these parity transformations.

Charge Conjugation

Charge conjugation is implemented in a way similar to the way it is implemented in conventional quantum field theory. In particular

$$\mathscr{C}X^\mu(\mathbf{y}, y^0)\mathscr{C}^{-1} = X^\mu(\mathbf{y}, y^0) \qquad (4.44)$$

Time Reversal

Since $X^0 = y^0$ in the Y Coulomb gauge in two-tier quantum theory the only non-trivial form of time reversal transformation $\mathscr{T}$ is based on $y^0 = -y^0$. This time reversal transformation is similar in part to to the conventional time reversal transformation in conventional quantum field theory. Therefore we will define $\mathscr{T}$ as the product of the operation of taking the complex conjugate of all c-numbers times a unitary operator $\mathscr{U}_Y$. Under $\mathscr{T}$ a scalar quantum field operator ϕ transforms as

$$\mathscr{T}\phi(\mathbf{X}(\mathbf{y}, y^0), X^0(\mathbf{y}, y^0))\mathscr{T}^{-1} = \phi(\mathbf{X}(\mathbf{y}, -y^0), X^0(\mathbf{y}, -y^0)) \qquad (4.45)$$

From the form of in ϕ eq. (4.2) we see that

$$\mathscr{T}a(\mathbf{p}, p^0)\mathscr{T}^{-1} = a(-\mathbf{p}, p^0) \qquad (4.46)$$

$$\mathscr{T}a^\dagger(\mathbf{p}, p^0)\mathscr{T}^{-1} = a^\dagger(-\mathbf{p}, p^0) \qquad (4.47)$$

$$\mathscr{T}X^i(\mathbf{y}, y^0)\mathscr{T}^{-1} = X^i(\mathbf{y}, -y^0) \qquad (4.48)$$

$$\mathscr{T}Y^i(\mathbf{y}, y^0)\mathscr{T}^{-1} = -Y^i(\mathbf{y}, -y^0) \qquad (4.49a)$$

$$\mathscr{T}a(\mathbf{k}, k^0,1)\mathscr{T}^{-1} = a(-\mathbf{k}, k^0,1) \qquad (4.49b)$$

$$\mathscr{T}a(\mathbf{k}, k^0,2)\mathscr{T}^{-1} = -a(-\mathbf{k}, k^0,2) \qquad (4.49c)$$

where i = 1,2,3 and assuming the convention: $\varepsilon(\mathbf{k},k^0,1)=-\varepsilon(-\mathbf{k},k^0,1)$ and $\varepsilon(\mathbf{k},k^0,2)=+\varepsilon(-\mathbf{k},k^0,2)$.

The unitary operator $\mathscr{U}_y$ is given by

$$\mathscr{U}_X = \exp\{-i\pi \int d^3p\ [a^\dagger(\mathbf{p}, p^0)a(\mathbf{p}, p^0) - a^\dagger(\mathbf{p}, p^0)a(-\mathbf{p}, p^0)]/2\} \quad (4.50a)$$

and

$$\mathscr{U}_y = \mathscr{U}_X \exp\{-i\pi \int d^3k\ [\sum_{\lambda=1}^{2} a^\dagger(\mathbf{k},k^0,\lambda)a(\mathbf{k},k^0,\lambda) - a^\dagger(\mathbf{k}, k^0, 1)a(-\mathbf{k},k^0,1) + a^\dagger(\mathbf{k}, k^0, 2)a(-\mathbf{k},k^0,2)]/2\} \quad (4.50b)$$

The separable Klein-Gordon Lagrangian (eq. 2.96) is invariant under our definition of time reversal.

We note

$$\mathscr{U}_y = \mathscr{P}_y \quad (4.50c)$$

Although the present theory is somewhat more complicated than conventional quantum field theory the overall nature of the $\mathscr{P}$, $\mathscr{C}$, and $\mathscr{T}$ transformations is the same.

5. Interacting Quantum Field Theory

Introduction

The form of quantum field theory that we have developed in chapter 3 can be used as the basis for new formulations of QED, Electroweak Theory, the Standard Model and a finite, unified theory of all the known interactions. The development of these theories requires a number of topics be addressed. This chapter covers perturbation theory. As much as possible, we attempt to retain the features of the standard approach so that the reader will more readily follow the discussion and more readily accept this new formalism. Originality is secondary to reality.

Y – Space – Quantum Dimensions™

There are many theories that posit additional dimensions beyond the observed dimensions of space and time.

The theories that we will develop are based on a quantum complexification of real space-time. These theories have the normal, real four dimensions of space and time found in everyday experience—both in the world around us, and at today's "high energy" accelerators. Today's energies are very much below the energies at which the additional imaginary, Quantum Dimensions™ become evident.

At ultrahigh energies, which may be of the order of the Planck mass (1.22×10^{19} GeV/c^2), the complexification of space-time becomes evident. The imaginary part of the complex four-dimensional space-time is implemented as Quantum Dimensions™ – the fluctuations of a quantum coodinate field Y^{μ}.

These quantum fluctuations serve to eliminate the divergences normally seen in quantum field theories. The result is a broad class of finite, two-tier quantum field theories, which include versions of the Standard Model and Quantum Gravity. Thus we can create a finite unified theory of all the interactions including gravity. And the two-tier formulation allows us to create quantum field theories with additional interactions (such as derivative couplings) that have hitherto been regarded as non-renormalizable. The two-tier formalism subdues these divergences, and produces theories that are finite to all orders in perturbation theory. Thus two-tier quantum field theory enables us to vastly broaden the possible forms of theories of Nature and provides a much more flexible framework for the development of quantum field theories. Yet at "low energies" a two-tier quantum field theory has the excellent feature of closely agreeing with the corresponding conventional quantum field theory. Thus we can develop a finite, two-tier theory such as two-tier QED or two-tier Electroweak theory that agrees (arbitrarily closely) with the corresponding conventional theory at experimentally accessible energies, and yet is well behaved at very high energies. Two-tier theories, properly formulated, implement a Correspondence Principle that accommodates the extremely accurate predictions of Quantum Electrodynamics, the predictions of Electroweak Theory, and the successes of the Standard Model.

The imaginary part of these complex coordinates is quantum in nature but is nevertheless physically significant. By choosing the Coulomb gauge for the Y field we continue to have a purely real time dimension while having complex spatial coordinates. Other choices of gauge are possible including an indefinite metric gauge (developed in QED by Gupta and Bleuler) that have complex space *and time* dimensions, and display manifest Lorentz invariance.

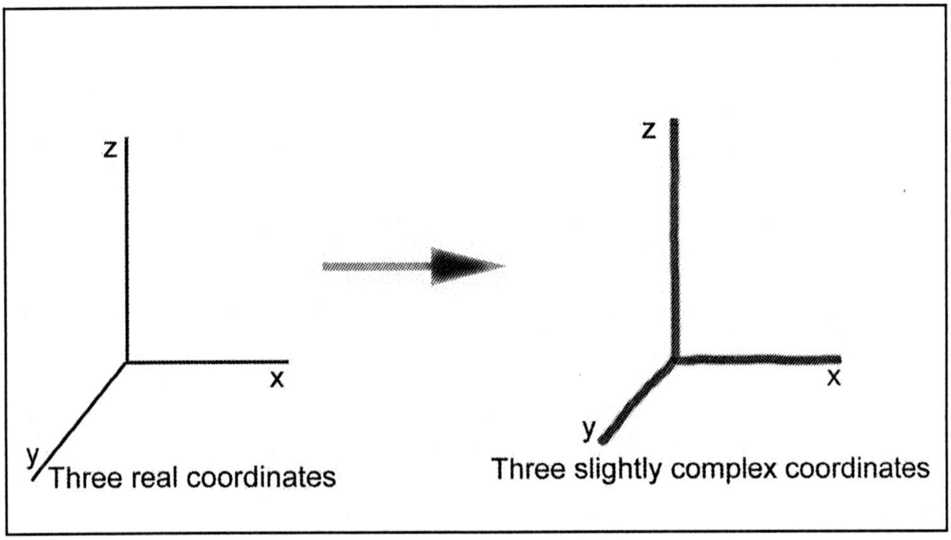

Figure 5.1. The change from purely real space to a slightly complex space with imaginary quantum fluctuations for each spatial axis in the Coulomb gauge of the Y field.

An Auxiliary Asymptotic Field

The definition of the asymptotic "free" in and out states is an issue in two-tier quantum field theory because the "free particle field" of the theory $\phi(X(y))$ is a "dressed" particle, ab initio, since it is cloaked in a cloud of Y particles as discussed in the passage following eq. 4.27.

While one could use it directly to define in and out asymptotic states it is more convenient initially to introduce a "fictitious" auxiliary asymptotic quantum field $\Phi(y)$ that will represent the equally fictitious "bare ϕ particle" in and out states.

We consider the case of a scalar field. Following the approach that quantum field theory traditionally uses, we define a scalar Klein-Gordon particle field with the physical mass m of the physical $\phi(X(y))$ particle.

$$\Phi(y) = \int d^3p\ N_m(p)\ [a(p)\ e^{-ip.y} + a^\dagger(p)\ e^{ip.y}] \tag{5.1}$$

using the raising and lowering operators of $\phi(X(y))$ (in eq. 4.2). The set of particle states of the $\Phi(y)$ has the familiar Fock space form

$$|\ p_1, p_2, \ldots p_n> = a^\dagger(p_1,)a^\dagger(p_2) \ldots a^\dagger(p_n)|0> \tag{5.2}$$

with powers of raising operators allowed since Φ particles are bosons. The set of particle states constitutes a complete orthonormal set of states. The corresponding bra states are defined by hermitean conjugation:

$$<p_1, p_2, \ldots p_n| = (|\ p_1, p_2, \ldots p_n>)^\dagger \tag{5.3}$$

We note that the energy spectrum of these states is positive definite with the hamiltonian

$$H_\Phi = P_\Phi{}^0 = \int d^3y\ \tfrac{1}{2}\ [\pi_\Phi{}^2 + (\partial\Phi/\partial X^i)^2 + m^2\Phi^2] \tag{5.4a}$$

$$= \int d^3p\ (\mathbf{p}^2 + m^2)^{\frac{1}{2}} a^\dagger(p)a(p) \tag{5.4b}$$

and momentum vector:

$$\mathbf{P}_\Phi = \int d^3p\ \mathbf{p}\ a^\dagger(p)a(p) \tag{5.5}$$

We will use this set of energy-momentum eigenstates to define asymptotic "in" and "out" states in perturbation theory.

Transformation Between $\Phi(y)$ and $\phi(X(y))$

For later use in the definition of the perturbation theory expansion we will determine the transformations between the in and out $\Phi(y)$ fields, and the in and out $\phi(X(y))$ fields. Let us define transformations $W_a(y)$ that transforms in and out $\Phi(y)$ fields to in and out $\phi(X(y))$ fields respectively:

$$\phi_a(X(y)) = W_a(y)\Phi_a(y)W_a^{-1}(y) \tag{5.6}$$

where the label a = "in" or a = "out" and where

$$\Phi_{in}(y) = \int d^3p\ N_m(p)\ [a_{in}(p)\ e^{-ip.y} + a_{in}^\dagger(p)\ e^{ip.y}] \tag{5.7}$$

$$\Phi_{out}(y) = \int d^3p\ N_m(p)\ [a_{out}(p)\ e^{-ip.y} + a_{out}^\dagger(p)\ e^{ip.y}] \tag{5.8}$$

$$\phi_{in}(X) = \int d^3p\ N_m(p)\ [a_{in}(p)\ {:}e^{-ip.(y + iY/M_c^2)}{:} + a_{in}^\dagger(p)\ {:}e^{ip.(y + iY/M_c^2)}{:}] \tag{5.9}$$

$$\phi_{out}(X) = \int d^3p\ N_m(p)\ [a_{out}(p)\ {:}e^{-ip.(y + iY/M_c^2)}{:} + a_{out}^\dagger(p)\ {:}e^{ip.(y + iY/M_c^2)}{:}] \tag{5.10}$$

In the Coulomb gauge of Y it is easy to show that

$$W_a(y) = \exp(\mathbf{Y}(y)\cdot\mathbf{P}_{\Phi a}/M_c^2) \tag{5.11}$$

and

$$W_a^{-1}(y) = \exp(-\mathbf{Y}(y)\cdot\mathbf{P}_{\Phi a}/M_c^2) \tag{5.12}$$

where the label a = "in" or a = "out", where the inner products in the exponentials are the usual spatial vector inner product, and where

$$\mathbf{P}_{\Phi a} = -\int d^3y\ \partial\Phi_a(y)/\partial y^0\ \nabla\Phi_a(y) = \int d^3p\ \mathbf{p}\ a_a^\dagger(p)a_a(p) \tag{5.12a}$$

is a spatial vector (the Φ spatial momentum operator) that is written solely in terms of $\Phi_a(y)$'s raising and lowering operators.

In addition to performing the transformation in eq. 5.6 $W_a(y)$ also performs a "translation" in Y^μ:

$$W_a(y)Y^i(y')W_a^{-1}(y) = Y^i(y') - i\Delta^{trij}(y' - y)P_{\Phi a}^j/M_c^2 \tag{5.13a}$$

where

$$i\Delta^{trij}(y' - y) = \int d^3k\ (e^{-ik\bullet(y' - y)} - e^{ik\bullet(y' - y)})(\delta_{jk} - k_jk_k/\mathbf{k}^2)/[(2\pi)^3 2\omega_k] \tag{5.13b}$$

We note that $W_a(y)$ is not a unitary operator but it is pseudo-unitary:

$$W_a^{-1}(y) = V\, W_a^{\dagger}(y)\, V^{-1} \tag{5.14}$$

where

$$V = \exp(-i\pi \sum_{\lambda=1}^{2} \int d^3k\, a^{\dagger}(k, \lambda)a(k, \lambda)) \tag{5.15}$$

is a unitary operator with the property

$$V\, Y^j(y)\, V^{-1} = -Y^j(y) \tag{5.16}$$

for j = 1,2,3. We note

$$V^{\dagger} = V^{-1} = V \tag{5.17}$$

and thus

$$V^2 = \mathrm{I} \tag{5.18}$$

V will be shown to be a metric operator in the following discussion.[21] We note that the Y particle (hermitean) number operator appears in eq. 5.9 in the expression for V:

$$N_Y = \sum_{\lambda=1}^{2} \int d^3k\, a^{\dagger}(k, \lambda)a(k, \lambda) \tag{5.19}$$

Thus states with an even number of Y particles have a V eigenvalue of one, and states with an odd number of Y particles have a V eigenvalue of minus one.

Model Lagrangian with ϕ^4 Interaction

We will develop perturbation theory using a scalar Lagrangian model with a ϕ^4 interaction term:

$$\mathscr{L}_S = J\mathscr{L}_F + \mathscr{L}_C \tag{5.20}$$

with

---

[21] P. A. M. Dirac, Proc. R. Soc. London A **180**, 1 (1942); T. D. Lee and G. C. Wick, Nucl. Phys. **B9**, 209 (1969); C. M. Bender, D. C. Brody and H. F. Jones, "Complex Extension of Quantum Mechanics" Phys. Rev. Letters **89**, 270401-1 (2002) and references therein.

$$\mathscr{L}_F = ½ [(\partial\phi/\partial X^\nu)^2 - m^2\phi^2] + \mathscr{L}_{Fint} \tag{5.21}$$

and

$$\mathscr{L}_C = -¼ F_Y{}^{\mu\nu}F_{Y\mu\nu} \tag{5.22}$$

with

$$F_{Y\mu\nu} = \partial Y_\mu/\partial y^\nu - \partial Y_\nu/\partial y^\mu \tag{5.23}$$

and

$$\mathscr{L}_{Fint} = 1/4!\ \chi_0\ \phi(X(y))^4 + ½ (m^2 - m_0{}^2)\phi^2 \tag{5.24}$$

where J is the Jacobian (as in chapter 2) and χ_0 is the bare coupling constant.

The conserved momentum operator is:

$$P_{F\beta} = \int d^3X\ \mathscr{T}_{F0\beta} \tag{5.25}$$

where

$$\mathscr{T}_{F\mu\nu} = - g_{\mu\nu}\mathscr{L}_F + \partial\mathscr{L}_F/\partial(\partial\phi/\partial X_\mu)\ \partial\phi/\partial X^\nu \tag{5.26}$$

is the ϕ field energy-momentum tensor with conservation law (eq. 2.110):

$$\partial P_{F\beta}/\partial X^0 = 0 \tag{5.27}$$

due to eq. 2.108.

The hamiltonian density (eq. 2.83) is

$$\mathscr{H}_F = \mathscr{T}_{F0\beta} = \mathscr{H}_{F0} + \mathscr{H}_{Fint} \tag{5.28}$$

with

$$\mathscr{H}_{F0} = ½ [\pi_\phi{}^2 + (\partial\phi/\partial X^i)^2 + m^2\phi^2] \tag{5.29}$$

$$\mathscr{H}_{Fint} = - 1/4!\ \chi_0\ \phi(X(y))^4 + ½ (m^2 - m_0{}^2)\phi(X(y))^2 \tag{5.30}$$

In-states and Out-States

In this section we will develop properties of in-fields and out-fields. We will use a somewhat more complicated procedure to set up the perturbation theory for the S matrix due to the introduction of imaginary coordinates. The procedure can be schematized as:

$$\Phi_{in}(y) \Rightarrow \phi_{in}(X(y)) \Rightarrow \phi(X(y)) \Rightarrow \phi_{out}(X(y)) \Rightarrow \Phi_{out}(y) \qquad (5.31)$$

In-states are constructed using the auxiliary field Φ_{in} which are then effectively restated in terms of $\phi_{in}(X(y))$ in order to make contact with our Lagrangian formalism. Then the interacting field $\phi(X(y))$ appears from $\phi_{in}(X(y))$ as part of a limiting ($y^0 \to -\infty$) process. Similarly out-states are constructed using the auxiliary field Φ_{out} which are then effectively restated in terms of $\phi_{out}(X(y))$. Again the interacting field $\phi(X(y))$ is related to $\phi_{out}(X(y))$ using the LSZ limiting process ($y^0 \to +\infty$).

Since much of the development differs only trivially from the standard treatment in textbooks we will simply "list" some relevant equations and let the reader pursue them further in quantum field theory textbooks.

ϕ In-Field

In order to define a perturbation theory for particle scattering we will consider aspects of the in-field $\phi_{in}(X(y))$ as well as in-field states – the field and states representing physical particles as $X^0 = y^0 \to -\infty$.

A. The in-field $\phi_{in}(X(y))$ satisfies the Klein-Gordon equation in the X variable:

$$(\Box_X + m^2)\, \phi_{in}(X) = 0 \qquad (5.32)$$

where

$$\Box_X = (\partial/\partial X^\nu)(\partial/\partial X_\nu)$$

B. Under coordinate displacements and Lorentz transformations $\Phi_{in}(y)$, $\phi_{in}(X(y))$, and $\phi(X(y))$ transform in the same way:

$$[P^\mu, \Phi_{in}(y)] = -\,i\partial\Phi_{in}/\partial y_\mu \qquad (5.33a)$$

$$[P^\mu, \phi_{in}(X)] = -\,i\partial\phi_{in}/\partial y_\mu \qquad (5.33b)$$

$$[P^\mu, \phi(X)] = -\,i\partial\phi/\partial y_\mu \qquad (5.34)$$

with the energy-momentum vector P^μ specified by eq. 2.57.

C. We can relate the asymptotic in-field $\phi_{in}(X(y))$ to the interacting field $\phi(X(y))$ using the equation of motion of $\phi(X(y))$

$$(\Box_X + m^2)\,\phi(X) = j(X) \tag{5.35}$$

where j(X) embodies the interaction. Using the physical mass m we find

$$(\Box_X + m^2)\,\phi(X) = j(X) + (m^2 - m_0{}^2)\phi(X) = j_{tot}(X) \tag{5.36}$$

If the current is taken to be the source of the scattered waves we may write

$$\sqrt{Z}\,\phi_{in}(X(y)) = \phi(X(y)) - \int d^4X(y')\,\Delta_{ret}(y - y')\,j_{tot}(X(y')) \tag{5.37}$$

$$= \phi(X(y)) - \int d^4y'\,J\,\Delta_{ret}(y - y')\,j_{tot}(X(y')) \tag{5.38}$$

where Z is a wave function renormalization constant, J is the Jacobian, and Δ_{ret} is a retarded Green's function.

D. We can define Φ_{in} in-field states with expressions like

$$|\,p_1, p_2, \ldots p_n\ in\rangle = a_{in}{}^\dagger(p_1)a_{in}{}^\dagger(p_2)\ldots a_{in}{}^\dagger(p_n)|0\rangle \tag{5.39}$$

with powers of raising operators allowed since Φ_{in} is a boson field. The set of all particle states constitutes a complete orthonormal set of states. The corresponding bra states are defined by hermitean conjugation:

$$\langle p_1, p_2, \ldots p_n\ in| = (|\,p_1, p_2, \ldots p_n\ in\rangle)^\dagger \tag{5.40}$$

ϕ Out-Field

In order to define a perturbation theory for particle scattering we will examine aspects of the out-field $\phi_{out}(X(y))$ as well as out-field states – the field and states representing physical particles as $X^0 = y^0 \rightarrow -\infty$.

A. The out-field $\phi_{out}(X(y))$ satisfies the Klein-Gordon equation in the X variable:

$$(\Box_X + m^2)\,\phi_{out}(X) = 0 \tag{5.41}$$

where

$$\Box_X = (\partial/\partial X^\nu)(\partial/\partial X_\nu)$$

B. Under coordinate displacements and Lorentz transformations $\Phi_{out}(y)$, $\phi_{out}(X(y))$, and $\phi(X(y))$ transform in the same way:

$$[P^\mu, \Phi_{out}(y)] = -\, i\partial\Phi_{out}/\partial y_\mu \tag{5.42a}$$

$$[P^\mu, \phi_{out}(X)] = -\, i\partial\phi_{out}/\partial y_\mu \tag{5.42b}$$

$$[P^\mu, \phi(X)] = -\, i\partial\phi/\partial y_\mu \tag{5.43}$$

with the energy-momentum vector P^μ specified by eq. 2.57.

C. We can relate the asymptotic out-field $\phi_{out}(X(y))$ to the interacting field $\phi(X(y))$ using the equation of motion of $\phi(X(y))$ specified by eq. 5.36:

$$\sqrt{Z}\, \phi_{out}(X(y)) = \phi(X(y)) - \int d^4X(y')\, \Delta_{adv}(y - y')\, j_{tot}(X(y')) \tag{5.44}$$

$$= \phi(X(y)) - \int d^4y'\, J\, \Delta_{adv}(y - y')\, j_{tot}(X(y')) \tag{5.45}$$

where Z is a wave function renormalization constant, J is the Jacobian, and Δ_{adv} is an advanced Green's function.

D. We can define Φ_{out} out-field states with expressions like

$$|\, p_1, p_2, \ldots\, p_n \text{ out}\rangle = a_{out}^\dagger(p_1,)a_{\Phi_{out}}^\dagger(p_2) \ldots\, a_{\Phi_{out}}^\dagger(p_n)|0\rangle \tag{5.46}$$

with powers of raising operators allowed since Φ_{out} is a boson field. The set of all particle states constitutes a complete orthonormal set of states. The corresponding bra states are defined by hermitean conjugation:

$$\langle p_1, p_2, \ldots\, p_n \text{ out}| = (|\, p_1, p_2, \ldots\, p_n \text{ out}\rangle)^\dagger \tag{5.47}$$

The Y Field

The Y field in the present model Lagrangian (eq. 5.20) is a free field and thus:

$$Y_{in}(y) = Y_{out}(y) = Y(y) \tag{5.48}$$

The states of the Y field have two general forms: 1.) States in a Fock space consisting of particle states that are eigenstates of the Y particle number operator (eq. 5.19); and 2.) Coherent states in a non-Fock space of generalized coherent states in an infinite tensor product space.[22]

The coherent ket states that arise in two-tier quantum field theory have the general form:

$$|y,\ p> = e^{-p.Y^{-}(y)/M_c^2}|0> \qquad (3.41)$$

as can be seen from an examination of $\phi_{in}(X(y))$. The corresponding bra state is:

$$<y,\ p| = (V|\ y,\ p>)^{\dagger} = <0|e^{+p.Y^{+}(y)/M_c^2} \qquad (5.49)$$

with V, the metric operator, reversing the sign of Y in the exponential. The inner product of coherent states is:

$$<y_1\ ,\ p_1|\ y_2,\ p_2> = \exp[-\ p_1{}^i p_2{}^j \Delta_{Tij}(y_1 - y_2)/M_c{}^4] \qquad (5.50)$$

showing the set of coherent states is not orthonormal and, in fact, is overcomplete. Comparing eq. 5.50 to eq. 4.12 gives $<y_1\ ,\ p|\ y_2,\ p> = R(p,\ y_1 - y_2)$.

The completeness of the set of states for each time y^0 can be verified by examining the projection operator:

$$\mathscr{R}_Y(y^0) = \because \exp[-i \int d^3y\ Y^{-}{}_i(y)|0><0|\pi^{+j}(y)] \because \qquad (5.51)$$

where

$$\pi^{+j}(y) = -\partial Y^{+j}(y)/\partial y^0 \qquad (5.52)$$

and where $\because$ represents an extended normal ordering operator:

$$\because \ \ldots \ \because$$

which is defined as placing creation operators to the left, projection operators in the center, and annihilation operators to the right. Thus eq 5.51 can be written

[22] See Kibble and other references on coherent states.

$$\mathscr{R}_Y = \sum_n (-i)^n \int d^3y_1 \ldots \int d^3y_n Y^{-j_1}(y_1)Y^{-j_2}(y_2)\ldots Y^{-j_n}(y_n)|0\rangle\langle 0|\pi^+_{j_1}(y_1)\pi^+_{j_2}(y_2)\ldots\pi^+_{j_n}(y_n) \tag{5.53}$$

where we note we have used the fact that |0><0| is a projection operator, and reduced |0><0| |0><0| … |0><0| to |0><0| in eq. 5.53.

We note

$$\mathscr{R}_Y (y^0)|\mathbf{y},y^0\ p\rangle = |\mathbf{y},y^0\ p\rangle \tag{5.54}$$

using eq. 3.22 and $\int d^3y_2\ p^i\ \Delta^{tr}{}_{ij}(y_1 - y_2)Y^{+j}(y_2) = \mathbf{p}\cdot\mathbf{Y}^+(y_1)$. Also

$$\mathscr{R}_Y(y^0)|n\rangle = |n\rangle \tag{5.55}$$

where |n> is any Y particle Fock state of finite particle number. In view of eqs. 5.54 and 5.55 $\mathscr{R}_Y$ is the identity operator on the Fock space and the space of generalized coherent Y field states. Thus the set of Y coherent states forms an overcomplete set of states. We will define the S matrix for any combination of Φ Fock space states and coherent Y states. The $\mathscr{R}_Y$ operator can be generalized to include Φ Fock space states:

$$\mathscr{R}_{\Phi Y}(y^0) = \because \exp[-i \int d^3y\ Y^-_j(y)\mathscr{R}_\Phi \pi^{+j}(y)] \because \tag{5.56}$$

where

$$\mathscr{R}_\Phi = \sum_n |n\rangle\langle n| \tag{5.57}$$

is a sum over all Φ Fock space states. Since $\mathscr{R}_\Phi$ is a projection:

$$[\mathscr{R}_\Phi]^N = \mathscr{R}_\Phi$$

for any power N, we find:

$$\mathscr{R}_{\Phi Y}(y^0) = \sum_n (-i)^n \int d^3y_1 \ldots \int d^3y_n Y^{-j_1}(y_1)Y^{-j_2}(y_2)\ldots Y^{-j_n}(y_n)\mathscr{R}_\Phi \pi^+_{j_1}(y_1)\pi^+_{j_2}(y_2)\ldots\pi^+_{j_n}(y_n) \tag{5.58}$$

As a result we have

$$\mathscr{R}_{\Phi Y}(y^0)\ |y,\ p;\ n_\Phi> = |y,\ p;\ n_\Phi> \tag{5.59}$$

for any combination of Y coherent states and Φ Fock space states n_Φ. Also

$$\mathscr{R}_{\Phi Y}(y^0)|n_\Phi> = |n_\Phi> \tag{5.60}$$

Thus $\mathscr{R}_{\Phi Y}$ is the identity operator on this space – the (over)complete space of the in and out states which we will use to define the S matrix of the scalar field theory specified by the Lagrangian eq. 5.20.

S Matrix

Following the standard definition of the S matrix we have:

$$S_{\alpha\beta} = <\alpha\ \text{out}|\beta\ \text{in}> \tag{5.61}$$

$$= <\alpha\ \text{in}|S|\beta\ \text{in}> \tag{5.62}$$

$$|0> = |0\ \text{in}> = |0\ \text{out}> = S|0\ \text{in}> \tag{5.63}$$

$$\Phi_{\text{in}}(y) = S\Phi_{\text{out}}(y)S^{-1} \tag{5.64}$$

and the other standard properties of the S matrix with the sole exception being the form of the unitarity relation (discussed later).

LSZ Reduction for Scalar Fields

In this section we will determine the reduction formula for the S matrix for scalar ϕ fields. Consider the S matrix element corresponding to an in state of particles β plus one ϕ particle of momentum p, and an out state α:

$$S_{\alpha\beta p} = <\alpha\ \text{out}|\beta p\ \text{in}> \tag{5.65}$$

After standard manipulations we have:

$$S_{\alpha\beta p} = \langle\alpha - p\ \text{out}|\beta\ \text{in}\rangle - i\langle\alpha\ \text{out}| \int d^3y\, f_p(y)\, \overleftrightarrow{\partial_0}\, [\Phi_{in}(y) - \Phi_{out}(y)]\, |\beta\ \text{in}\rangle \tag{5.66}$$

where $\langle\alpha - p\ \text{out}|$ is an out state with a particle of momentum p removed (if present) and where

$$f(y^0)\, \overleftrightarrow{\partial_0}\, g(y^0) = f(y^0)\, \partial g(y^0)/\partial y^0 - \partial f(y^0)/\partial y^0\, g(y^0) \tag{5.67}$$

and

$$f_p(y) = N_m(p)e^{-ip\cdot y} \tag{5.68}$$

with $N_m(p)$ specified by eq. 3.6.
We now express

$$S_{\alpha\beta p} = S_{\alpha-p\beta} - i\langle\alpha\ \text{out}| \int d^3y\, f_p(y)\, \overleftrightarrow{\partial_0}\, W^{-1}[\phi_{in}(X(y)) - \phi_{out}(X(y))]W|\beta\ \text{in}\rangle \tag{5.69}$$

using $W(y) = W_{in}(y)$ with

$$\Phi_a(y) = W_a^{-1}(y)\phi_a(X(y))W_a(y) \tag{5.70}$$

where the label a = “in” or a = “out”, and where

$$W_a(y) = \exp(\mathbf{Y}(y)\cdot\mathbf{P}_{\Phi a}/M_c^2) \tag{5.71}$$

and

$$W_a^{-1}(y) = \exp(-\mathbf{Y}(y)\cdot\mathbf{P}_{\Phi a}/M_c^2) \tag{5.72}$$

in the Coulomb gauge of Y with $\mathbf{P}_{\Phi a}$ the momentum spatial vector defined by eq. 5.12a.

We note that the interacting $\phi(X(y))$ approaches the in and out fields $\phi_{in}(X(y))$ and $\phi_{out}(X(y))$ in the limit that $y^0 \to -\infty$ and $y^0 \to +\infty$ respectively in the sense of Lehmann, Symanzik and Zimmermann[23] which we *symbolize* as:

$$\phi(X(y)) \to \sqrt{Z}\, \phi_{in}(X(y)) \quad \text{as} \quad y^0 \to -\infty \tag{5.73}$$

---

[23] H. Lehmann, K. Symanzik and W. Zimmermann, Nuov. Cim., **1**, 1425 (1955); W. Zimmermann, Nuov. Cim., **10**, 567 (1958); O. W. Greenberg, Doctoral Dissertation, Princeton University 1956.

$$\phi(X(y)) \rightarrow \sqrt{Z}\, \phi_{out}(X(y)) \quad \text{as} \quad y^0 \rightarrow +\infty \tag{5.74}$$

with $\sqrt{Z}$ defined in eqs. 5.37 and 5.44. Thus we can rewrite eq. 5.69 as

$$S_{\alpha\beta p} = S_{\alpha-p\beta} + iZ^{-1/2} \left(\lim_{y^0\rightarrow+\infty} - \lim_{y^0\rightarrow-\infty}\right) \langle\alpha\ \text{out}|\int d^3y\, f_p(y)\, \overleftrightarrow{\partial_0}\, W^{-1}\phi(X(y))W|\beta\ \text{in}\rangle \tag{5.75}$$

which after standard manipulations becomes

$$S_{\alpha\beta p} = S_{\alpha-p\beta} + iZ^{-1/2} \int d^4y\, f_p(y)(\Box_y + m^2)\langle\alpha\ \text{out}|\, W(y)^{-1}\phi(X(y))W(y)|\beta\ \text{in}\rangle \tag{5.76}$$

Eq. 5.76 is similar to the usual LSZ reduction formula except for the appearance of the W(y) operator and its inverse. We note that $W(y) = W_{in}(y)$ still because $\mathbf{P}_{\Phi in}$ is independent of y^0.

Similarly an out ϕ particle can be reduced from an S matrix part. For example,

$$\begin{aligned}\langle\alpha\ \text{out}|W^{-1}(y)\phi(X(y))W(y)|\beta\ \text{in}\rangle &= \langle\alpha-p'\ \text{out}|W^{-1}(y)\phi(X(y))W(y)|\beta-p'\ \text{in}\rangle \\ &- i\langle\alpha-p'\ \text{out}|\int d^3y'\, [W^{-1}(y')\phi_{in}(X(y'))W(y')W^{-1}(y)\phi(X(y))W(y) - \\ &- W^{-1}(y)\phi(X(y))W(y)W^{-1}(y')\phi_{out}(X(y'))W(y')]|\beta\ \text{in}\rangle\, \overleftrightarrow{\partial_0}\, f_{p'}^{*}(y')\end{aligned} \tag{5.77}$$

which becomes

$$\begin{aligned}\langle\alpha\ \text{out}|W^{-1}(y)\phi(X(y))W(y)|\beta\ \text{in}\rangle &= \langle\alpha-p'\ \text{out}|\varphi(y)|\beta-p'\ \text{in}\rangle + \\ &+ iZ^{-1/2}\int d^4y'\, \langle\alpha-p'\ \text{out}|T(\varphi(y')\varphi(y))|\beta\ \text{in}\rangle\, (\overleftarrow{\Box}_{y'} + m^2)\, f_{p'}^{*}(y')\end{aligned} \tag{5.78}$$

where the time ordered product is defined with respect to ordering by y^0 and where

$$\varphi(y) = W^{-1}(y)\phi(X(y))W(y) \tag{5.79}$$

These results directly generalize to multi-particle in and out states:

$$\langle p_1, p_2, \ldots p_n\ \text{out}|\, q_1, q_2, \ldots q_m\ \text{in}\rangle = \ldots \langle 0|T(\varphi(y'_1) \ldots \varphi(y'_n)\varphi(y_1) \ldots \varphi(y_m))|0\rangle \ldots \tag{5.80}$$

thus reducing the development of the perturbation theory of the S matrix to the evaluation of time ordered products such as

$$<0|T(\varphi(y_1) \ldots \varphi(y_n))|0> \tag{5.81}$$

Coherent in and out States

In addition to ϕ particle in and out states there will also be coherent Y in and out states. The Y field is a free field in two-tier quantum field theories as discussed in this book. It *could* assume a new role as an interacting field in a modified two-tier quantum gravity.

Since the Y field is free, its role in the development and perturbative calculation of S matrix elements is particularly simple. We begin by noting that conservation of momentum implies that we consider coherent states of fixed total momentum. To that end we define a set of coherent Y states with

$$|q\ C> = \int d^4y\ e^{-iq.X^{-}(y)}|0> = \int d^4y\ e^{-iq\cdot y - \mathbf{q}.\mathbf{Y}^{-}(y)/M_c^2}|0> \tag{5.82}$$

where C is for "Coherent." Expanding the exponential term $\mathbf{q}\cdot\mathbf{Y}^{-}(y)/M_C^2$ and performing the integration over y shows |q C> to be superposition of multi-particle Y states with each term in the superposition having total momentum q. Since the Y field is free no distinction need be made between in and out coherent states. We can define coherent bra states with

$$<C\ q| = (V|\ q\ C>)^{\dagger} = <0| \int d^4y\ e^{iq\cdot y + \mathbf{q}.\mathbf{Y}^{+}(y)/M_c^2} \tag{5.83}$$

where the effect of the V operator is to flip the sign of the Y^+ term. The role of these states will be seen to be irrelevant for the S matrix of two-tier quantum field theory if the Lagrangian interaction does not have an explicit dependence on the Y field.

6. Perturbation Theory

The U Matrix

The U matrix for a two-tier theory is developed in a way similar to conventional field theory starting from the defining relations:

$$\phi(X(y)) = U^{-1}\phi_{in}(X(y))U \qquad (6.1)$$

$$\pi_{\phi}(X(y)) = U^{-1}\pi_{\phi in}(X(y))U \qquad (6.2)$$

From eq. 5.29 we define the free field hamiltonian

$$H_{F0in}(\ \phi_{in},\ \pi_{\phi in}) = \int d^3X\ \mathscr{H}_{F0}(\ \phi_{in},\ \pi_{\phi in}) \qquad (6.3)$$

Noting $X^0 = y^0$ in the Y Coulomb gauge we find

$$\partial\phi_{in}/\partial y^0 = i[H_{F0in},\ \phi_{in}(X)] \qquad (6.4)$$

$$\partial\pi_{\phi in}/\partial y^0 = i[H_{F0in},\ \pi_{\phi in}(X)] \qquad (6.5)$$

For the entire hamiltonian (eq. 5.28) we have

$$\partial\phi/\partial y^0 = i[H_F,\ \phi(X)] \qquad (6.6)$$

$$\partial\pi_{\phi}/\partial y^0 = i[H_F,\ \pi_{\phi}(X)] \qquad (6.7)$$

with

$$H_F(\ \phi,\ \pi_{\phi}) = \int d^3X\ \mathscr{H}_F(\ \phi,\ \pi_{\phi}) \qquad (6.8)$$

Combining the above equations in the standard way yields a familiar differential equation for the U matrix:

$$i\partial U(y^0)/\partial y^0 = (H_{Fint} + E_0(t))U(y^0) \tag{6.9}$$

where $E_0(t)$ is a c-number function of y^0 that we can set equal to 0 (as it would be cancelled later anyway), and where

$$H_{Fint}(\phi_{in}, \pi_{\phi in}) = \int d^3X\, \mathscr{H}_{Fint}(\phi_{in}, \pi_{\phi in}) \tag{6.10}$$

with $\mathscr{H}_{Fint}$ given by eq. 5.30. Solving for U gives the familiar time ordered exponential:

$$U(y^0) = T\Big(\exp[-i\int_{-\infty}^{t} dy^0\, H_{Fint}]\Big) \tag{6.11a}$$

which is a symbolic notation for:

$$U(y^0) = 1 + \sum_{n=1}^{\infty} (-i)^{-n}(n!)^{-1}\int_{-\infty}^{y^0} dy_1^{\,0} \ldots \int_{-\infty}^{y^0} dy_n^{\,0}\, T(H_{Fint}(y_1^{\,0}) \ldots H_{Fint}(y_n^{\,0})) \tag{6.11b}$$

We note for later use that the hermiticity of H_{Fint} is not used in the derivation of eq. 6.11. Thus eq. 6.11 would still hold if H_{Fint} were not hermitean.

Reduction of Time Ordered φ Products

In the previous chapter we reduced the calculation of the S matrix to the evaluation of time ordered products of the form

$$\tau(y_1, \ldots, y_n) = \langle 0|T(\varphi(y_1) \ldots \varphi(y_n))|0\rangle \tag{6.12}$$

where $\varphi(y)$ is specified by eq. 5.79. Expanding the terms within eq. 6.12 we find

$$\varphi(y_1) \ldots \varphi(y_n) = W^{-1}(y_1)\phi(X(y_1))W(y_1)W^{-1}(y_2)\phi(X(y_2))W(y_2) \ldots W^{-1}(y_n)\phi(X(y_n))W(y_n) \tag{6.13}$$

which can re-expressed as

$$W^{-1}(y_1)U^{-1}(y_1{}^0)\phi_{in}(X(y_1))U(y_1{}^0)W(y_1)W^{-1}(y_2)U^{-1}(y_2{}^0)\phi_{in}(X(y_2))U(y_2{}^0)W(y_2)\ \ldots \tag{6.14}$$

using eq. 6.1 and denoting $W_{in}(y)$ as W(y). Defining

$$\mathscr{U}(y_1, y_2) = U(y_1{}^0)W(y_1)W^{-1}(y_2)U^{-1}(y_2{}^0) \tag{6.15}$$

we see eq. 6.14 can be rewritten as

$$W^{-1}(y_1)U^{-1}(y_1{}^0)\phi_{in}(X(y_1))\mathscr{U}(y_1, y_2)\phi_{in}(X(y_2))\mathscr{U}(y_2, y_3)\phi_{in}(X(y_3))\ \ldots\phi_{in}(X(y_n))U(y_n{}^0)W(y_n) \tag{6.16}$$

From eqs. 5.71 and 5.72

$$\mathscr{U}(y_1, y_2) = U(y_1{}^0)\exp((\mathbf{Y}(y_1) - \mathbf{Y}(y_2))\cdot\mathbf{P}_{\Phi a}/M_c{}^2)U^{-1}(y_2{}^0) \tag{6.17}$$

Defining

$$W(y_1, y_2) = \exp((\mathbf{Y}(y_1) - \mathbf{Y}(y_2))\cdot\mathbf{P}_{\Phi a}/M_c{}^2) \tag{6.18}$$

and looking ahead to the Wick expansion of the time ordered product of eq. 6.12 we note that the only time ordered products involving $W(y_1, y_2)$ that would appear in the expansion are

$$<0|T(\phi_{in}(X(y))W(y_1, y_2))|0> = 0 \tag{6.19}$$

$$<0|T(W(y_1, y_2)W(y_3, y_4))|0> = 1 \tag{6.20}$$

Thus the $W(y_1, y_2)$ factor in eq. 6.17 may be set to the value one with the result

$$\mathscr{U}(y_1, y_2) \equiv U(y_1{}^0)U^{-1}(y_2{}^0) = U(y_1{}^0, y_2{}^0) \tag{6.21}$$

$U(y_1{}^0, y_2{}^0)$ is the conventionally defined U matrix satisfying

$$i\partial\, U(y_1{}^0, y_2{}^0)/\partial y_1{}^0 = iH_{Fint}\, U(y_1{}^0, y_2{}^0) \tag{6.22}$$

with the boundary condition

$$U(y^0, y^0) = 1 \tag{6.23}$$

Then, paralleling the standard approach we find an expression for the U matrix:

$$U(y_1{}^0, y_2{}^0) = T\left(\exp[-i \int_{y_2{}^0}^{y_1{}^0} dy'^0 d^3X(y')\, \mathscr{H}_{Fint}(\phi_{in}(X(y')), \pi_{\phi in}(X(y')))]\right) \tag{6.24}$$

The $U(y_1{}^0, y_2{}^0)$ matrix satisfies the conventional multiplication rule:

$$U(y_1{}^0, y_3{}^0) = U(y_1{}^0, y_2{}^0)U(y_2{}^0, y_3{}^0) \tag{6.25}$$

The inverse of $U(y_1, y_2)$ is

$$U^{-1}(y_1{}^0, y_2{}^0) = U(y_2{}^0, y_1{}^0) \tag{6.26}$$

We now return to eq. 6.16, which can now be written in the form:

$$U^{-1}(y^0)U(y^0, y_1{}^0)\phi_{in}(X(y_1))U(y_1{}^0, y_2{}^0)\phi_{in}(X(y_2))U(y_2{}^0, y_3{}^0) \ldots \phi_{in}(X(y_n))U(y_n{}^0, -y^0)U(-y^0) \tag{6.27}$$

where y^0 is a reference time that is later than all other times, and $-y^0$ is earlier than all the other times, in the time-ordered product. As a result the time-ordered product in eq. 5.80 can be expressed in a symbolic notation as:

$$<0|U^{-1}(y^0)T(\phi_{in}(X(y_1))\phi_{in}(X(y_2)) \ldots \phi_{in}(X(y_n))U(y^0, -y^0))U(-y^0)|0> \tag{6.28}$$

The analysis of eq. 6.28 as $y^0 \to \infty$ follows the standard path, which begins by noting

$$U(-y)|0> = \lambda_-|0> \qquad \text{when } y^0 \to \infty \tag{6.29a}$$

$$U(y)|0> = \lambda_+|0> \qquad \text{when } y^0 \to \infty \tag{6.29b}$$

following a standard textbook proof, which, in turn, leads to:

$$\lambda_-\lambda_+^* = <0|T\Big(\exp[\,+i\int_{-\infty}^{\infty} dy'^0 d^3X(y')\,\mathscr{H}_{Fint}(\phi_{in}(X(y')),\ \pi_{\phi in}(X(y')))]\Big)|0> \tag{6.30}$$

$$= \Big[<0|T\Big(\exp[\,-i\int_{-\infty}^{\infty} dy'^0 d^3X(y')\,\mathscr{H}_{Fint}(\phi_{in}(X(y')),\ \pi_{\phi in}(X(y')))]\Big)|0>\Big]^{-1} \tag{6.31}$$

Thus the time ordered product of eq. 6.12, which appears in the evaluation of the S matrix element in eq. 5.80, can be symbolically written as:

$$\tau(y_1, \ldots, y_n) = \frac{<0|T(\phi_{in}(X(y_1)) \ldots \phi_{in}(X(y_n))U(\infty, -\infty))|0>}{<0|T\Big(\exp[-i\int dy'^0 d^3X(y')\mathscr{H}_{Fint}(\phi_{in}(X(y')),\pi_{\phi in}(X(y')))]\Big)|0>} \tag{6.32}$$

in the limit $y^0 \to \infty$.

The $\int d^3X$ Integration

The integration over the X space coordinates presents the difficulty of a functional integration of a q-number that needs to be properly defined. Since

$$X^\mu(y) = y^\mu + i\,Y^\mu(y)/M_c^2 \tag{3.12}$$

by definition and since, in the Y Coulomb gauge we have $X^0(y) = y^0$ due to $Y^0 = 0$, the classical Jacobian for the transformation from y to X coordinates is the absolute value of:

$$J = \varepsilon^{ijk}\left(\delta^{1i} + \frac{i}{M_C^2}\frac{\partial Y^1}{\partial y^i}\right)\left(\delta^{2j} + \frac{i}{M_C^2}\frac{\partial Y^2}{\partial y^j}\right)\left(\delta^{3k} + \frac{i}{M_C^2}\frac{\partial Y^3}{\partial y^k}\right) \tag{6.33}$$

The Jacobian appears in a change of integration variables:

$$\int d^3X = \int d^3y\, J \qquad (6.34)$$

and

$$\int d^4X = \int d^4y\, J \qquad (6.35)$$

in the Y Coulomb gauge.

The development of the change of variables for c-number coordinate transformations is well known. The situation changes when one set of coordinates are in fact q-numbers. The second quantization of the Y field requires the definition of J to be clarified since the product of fields at the same position is normally undefined. At the very least we therefore require J to be normal ordered to avoid infinities.

While normal ordering eliminates the infinities that would otherwise be present, J still presents a problem because it is still effectively an interaction term that appears as a factor even in the free field part of the hamiltonian. This situation appears to be unsatisfactory in the present scalar quantum field theory in which Y is not intended to play a direct dynamical role but rather a passive role as a coordinate. In other words the ϕ field is supposed to be the only interacting field. Thus it makes physical sense to also require that J be reduced to a c-number by requiring it to be a vacuum expectation value as well, which reduces it to the value of unity.

$$<0|:J:|0> = 1 \qquad (6.36)$$

In addition, it is important to note that the Jacobian is a feature of classical field theory and needs to be suitably implemented in the corresponding quantum field theory. Quantum field theory features are not mirror images of the corresponding classical field theory. There are well known differences between classical field theory and conventional quantum field theory such as the Adler-Bell-Jackiw anomaly. Thus we can choose $J = <0|:J:|0> = 1$ in the definition of the corresponding quantum field theory.

Lastly we note the Jacobian has the value unity up to order M_c^{-4} in the gauge where $\nabla\cdot\mathbf{Y} = 0$. Thus at current energies, which are very small in comparison to M_c, the Jacobian is unity to very good approximation. These issues are discussed later in chapter 8 in the section on path integral formulations of gauge theories where additional justification is provided for eq. 6.36.

As a result the time ordered product (eq. 6.32) becomes:

$$\tau(y_1,\ldots, y_n)= \frac{<0|T(\phi_{in}(X(y_1))\ldots\phi_{in}(X(y_n))\exp[-i\int d^4y'\mathscr{H}_{Fint}(\phi_{in}(X(y')))]) |0>}{<0|T\big(\exp [-i\int d^4y'\mathscr{H}_{Fint}(\phi_{in}(X(y')))]\big)|0>} \tag{6.37}$$

Coherent Y In and out states

The Y fields have no interactions and are thus free fields in the model Lagrangian under consideration and in the two-tier quantum field theories that we will construct later.

Coherent Y Field States

We will use coherent states of the type of eqs. 5.82 and 5.83 $|q\ C>$ consisting of superpositions of Y particles of fixed total momentum q:

$$|q\ C> = \int d^4y\ e^{-iq.X^-(y)}|0> = \int d^4y\ e^{-iq.y\ -\ q.Y^-(y)/M_c^2}|0> \tag{6.38a}$$

and

$$<C\ q| = (V|\ q\ C>)^\dagger = <0|\int d^4y\ e^{iq.y\ +\ q.Y^+(y)/M_c^2} \tag{6.38b}$$

In principle, in-states and out-states containing these coherent states are possible. However for Lagrangians which do not have interaction terms explicitly containing Y field factors the S matrix is "block diagonal" in the sense that if an in-state does not contain a Y particle, or a Y coherent stateC then the set of possible out-states will not contain Y particles or coherent states. The proof is based on the expansion of S matrix elements using Wick's theorem in products of time ordered products of pairs of in field operators. Since

$$<0|\ T(\phi_{in}(X(y_1))Y^j(y_2))|0> = 0 \tag{6.39}$$

and

$$<0|\ T(\phi_{in}(X(y_1))e^{-\ q.Y^-(y)/M_c^2})|0> = <0|\ T(\phi_{in}(X(y_1))e^{+q.Y^+(y)/M_c^2})|0> = 0 \tag{6.40}$$

S matrix elements with no incoming Y particle or coherent states will have zero matrix elements to produce outgoing Y particle or coherent states.

Unitarity

For many years it has been evident that modified field theories[11, 17, 24] might offer some hope of avoiding the divergences of conventional quantum field theory. Usually these theories suffer from unitarity problems: negative norms and negative probabilities. In the absence of a physically acceptable interpretation of negative probabilities, these theories have been thought to be unsatisfactory.

The two-tier type of quantum field theory *superficially* also appears to have a unitarity problem due to the non-hermitean nature of two-tier hamiltonians. The lack of hermiticity is due entirely to the appearance of iY^μ in the X^μ field coordinates.

Thus the interaction hamiltonian in eq. 6.37 is not hermitean:

$$H_{Fint} = \int d^3y' \mathscr{H}_{Fint}(\phi_{in}(\, y' + iY(y')/M_c^2)) \tag{6.41}$$

and

$$H_{Fint}^{\dagger} = \int d^3y' \mathscr{H}_{Fint}(\phi_{in}(\, y' - iY(y')/M_c^2)) \neq H_{Fint} \tag{6.42}$$

The relation between H_{Fint} and its hermitean conjugate is

$$H_{Fint} = V\, H_{Fint}^{\dagger}\, V \tag{6.43}$$

where $V^2 = I$ is the metric operator defined in eqs. 5.15 – 5.18. Thus the S matrix is not unitary; the S matrix is pseudo-unitary:

$$S^{-1} = V\, S^{\dagger}\, V \tag{6.44}$$

and

$$S^{\dagger} VS = V \tag{6.45}$$

We will now show that the S matrix is effectively *unitary between physical asymptotic states* which in the present scalar field theory are states consisting of ϕ particles only. As we noted in the discussion of coherent states a few pages ago, the S matrix is *block diagonal* in the

[24] S. Blaha, Phys.Rev. **D10**, 4268 (1974); S. Blaha, Phys.Rev. **D11**, 2921 (1975); S. Blaha, Nuovo Cim. **A49**, :113 (1979); S. Blaha, "Generalization of Weyl's Unified Theory to Encompass a Non-Abelian Internal Symmetry Group" SLAC-PUB-1799, Aug 1976; S. Blaha, "Quantum Gravity and Quark Confinement" Lett. Nuovo Cim. **18**, 60 (1977); S. Blaha, "The Local Definition of Asymptotic Particle States" Nuovo Cim. **A49**, 35 (1979) and references therein.

sense that in-states consisting only of ϕ particles will only produce out-states consisting of ϕ particles only. The consequence of block diagonality is that *S is unitary within the physical states sector*. To prove this point, consider eq 6.45 between physical states – each consisting of ϕ particles only.

$$<\text{f in}|V|\text{i in}> = <\text{f in}|S^\dagger VS|\text{i in}> = \sum_{n,\,m} <\text{f in}|S^\dagger|n><n|V|m><m|S|\text{i in}> \tag{6.46}$$

where the block diagonality of S (and the diagonality of V) limits the intermediate states |n> and |m> to states containing ϕ particles only. We note V has the eigenvalue one between states consisting of ϕ particles only. Thus eq. 6.46 becomes

$$\delta_{\text{fi}} = \sum_n <\text{f in}|S^\dagger|n><n|S|\text{i in}> \tag{6.47}$$

or

$$S_{\text{phys}}^\dagger S_{\text{phys}} = I$$

or

$$S_{\text{phys}}^\dagger = S_{\text{phys}}^{-1} \tag{6.48}$$

proving unitarity between the physical states – the states consisting of ϕ particles only. This proof goes through with no change for the two-tier version of the Standard Model with physical states consisting of any combination of particles in the Standard Model – but not including any Y particle states or Y coherent states. We note the Standard Model S matrix has the same sort of block diagonality as seen here. The S matrix is block diagonal between physical states (which are the states without Y particles or Y coherent states) and is unitary within that block.

Comparison with String and Superstring Theories

A pseudo-unitary S matrix (with some states of positive metric and others of negative metric) is a common feature of string and superstring theories. These theories, like two-tier theories, define a set of physical, positive metric states whose S matrix elements satisfy the unitarity requirement. Thus their S matrix has a form of block diagonality which enables them to have a "physical" S matrix in a manner analogous to two-tier quantum field theory.

Perturbation Expansion

Perturbation theory in two-tier quantum field theory is very similar to conventional perturbation theory. The difference is in the form of the propagators, which have a high energy damping factor R(p, z) that eliminates infinities that normally appear at high energy in conventional quantum field theories.

In order to develop a feeling for two-tier perturbation theory we will calculate a few low order terms in the perturbation theory of the model scalar ϕ^4 theory that we have been using as an example in this chapter.

Fig. 6.1 contains the lowest order diagram for the scattering of two ϕ particles into an out-state of two ϕ particles.

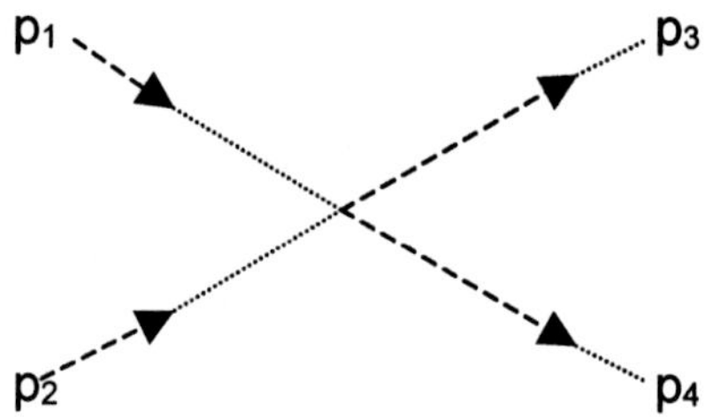

Figure 6.1. Lowest order quartic interaction diagram.

The S matrix element for this diagram is

$$S_1 = (iZ^{-\frac{1}{2}})^4 \left(\tfrac{1}{4!}\, i\mathcal{X}_0\right) \prod_{j=1}^{4} \int d^4y_j\, d^4y\, f_{p_1}(y_1) f_{p_2}(y_2) f_{p_3}^{*}(y_3) f_{p_4}^{*}(y_4)(\Box_{y_1} + m^2)\cdot$$

$$\cdot(\Box_{y_2}+m^2)(\Box_{y_3}+m^2)(\Box_{y_4}+m^2)\langle 0|T(\phi_{in}(X(y_1))\ldots\phi_{in}(X(y_4)):(\phi_{in}(X(y))^4:)|0\rangle \tag{6.49}$$

with $f_p(y)$ specified by eq. 5.68. Expanding the time ordered product and realizing there are 4! ways of combining the four field factors in the interaction hamiltonian leads to:

$$S_1 = (iZ^{-\frac{1}{2}})^4 \left(\tfrac{1}{4}\, i\mathcal{X}_0\right) \prod_{j=1}^{4} \int d^4y_j\, d^4y\, f_{p_1}(y_1) f_{p_2}(y_2) f_{p_3}^{*}(y_3) f_{p_4}^{*}(y_4)(\Box_{y_1} + m^2)\cdot$$

$$\cdot(\Box_{y_2}+m^2)(\Box_{y_3}+m^2)(\Box_{y_4}+m^2)\, i\Delta_F^{TT}(y_1-y)\, i\Delta_F^{TT}(y_2-y)\, i\Delta_F^{TT}(y_3-y)\, i\Delta_F^{TT}(y_4-y) \tag{6.50}$$

where

$$i\Delta_F^{TT}(y_1 - y_2) = <0|T(\phi(X(y_1)),\phi(X(y_2)))|0> \tag{6.51}$$

$$= i \int \frac{d^4p\, e^{-ip\bullet(y_1 - y_2)} R(\mathbf{p}, y_1 - y_2)}{(2\pi)^4 (p^2 - m^2 + i\varepsilon)} \tag{6.52}$$

with

$$R(\mathbf{p}, y_1 - y_2) = \exp[- p^i p^j \Delta_{Tij}(y_1 - y_2)/M_c^4] \tag{6.53}$$

(summations are over space indices only in the Y Coulomb gauge) and

$$\Delta_{Tij}(z) = \int d^3k\, e^{-ik\bullet z} (\delta_{ij} - k_i k_j/\mathbf{k}^2)/[(2\pi)^3 2\omega_k] \tag{6.54}$$

From chapter 4 we have:

$$R(\mathbf{p}, y_1 - y_2) = \exp\{ - p^2[A(v) + B(v)\cos^2\theta] / [4\pi^2 M_c^4 z^2] \} \tag{6.55}$$

with

$$z^\mu = y_1^\mu - y_2^\mu \tag{6.56}$$

$$z = |\mathbf{z}| = |\mathbf{y_1} - \mathbf{y_2}| \tag{6.57}$$

$$p = |\mathbf{p}| \tag{6.58}$$

$$v = |z^0| / z \tag{6.59}$$

$$A(v) = (1 - v^2)^{-1} + .5v \ln[(v - 1)/(v + 1)] \tag{6.60}$$

$$B(v) = v^2(1 - v^2)^{-1} - 1.5v \ln[(v - 1)/(v + 1)] \tag{6.61}$$

$$\mathbf{p}\cdot\mathbf{z} = pz \cos\theta \tag{6.62}$$

and with $|\mathbf{p}|$ denoting the length of the spatial vector $\mathbf{p}$, while $|z^0|$ is the absolute value of z^0.

Letting $y_i = w_i + y$ yields

$$S_1 = (iZ^{-½})^4(¼\, i\chi_0)(2\pi)^4\, \delta^4(p_3 + p_4 - p_1 - p_2)\, \mathcal{N}^+(p_4)\mathcal{N}^+(p_3)\mathcal{N}(p_2)\mathcal{N}(p_1) \tag{6.63}$$

where

$$\mathcal{N}(p) = i\int d^4w\, f_p(w)(\Box + m^2)\Delta_F^{TT}(w) \tag{6.64}$$

$$\mathcal{N}^+(p) = i\int d^4w\, f_p{}^*(w)(\Box + m^2)\Delta_F^{TT}(w) \tag{6.65}$$

are "normalizations" of the "external legs" – the in and out states due to the Y field cloud around each particle. In the limit of low momentum ($p \ll M_C$):

$$\mathcal{N}(p) = \mathcal{N}^+(p) \to -i\, [(2\pi)^3\, 2p^0\,]^{-½} \tag{6.66}$$

which the reader will note is the standard normalization factor for external scalar field legs in conventional quantum field theory.

Higher Order Diagram With a Loop

We will now consider the simplest one loop scattering diagrams in the scalar ϕ^4 theory.

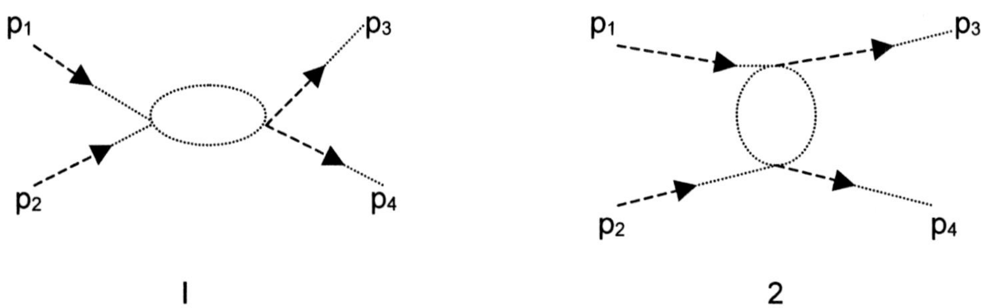

Figure 6.2. Lowest order loop scattering diagrams.

The S matrix element for these diagrams is contained in

$$S_2 = (iZ^{-1/2})^4(1/4!\ i\chi_0)^2 \prod_{j=1}^{4} \int d^4y_j\, d^4y'_1\, d^4y'_2\, f_{p_1}(y_1)f_{p_2}(y_2)f_{p_3}^{\ *}(y_3)f_{p_4}^{\ *}(y_4)\cdot$$
$$\cdot(\Box_{y_1}+m^2)(\Box_{y_2}+m^2)(\Box_{y_3}+m^2)(\Box_{y_4}+m^2)<0|T(\phi_{in}(X(y_1))\ldots\phi_{in}(X(y_4))$$
$$:(\phi_{in}(X(y'_1))^4::(\phi_{in}(X(y'_2))^4:)|0>/2! \tag{6.67}$$

together with some other disconnected diagrams.

Expanding the time ordered product and keeping only the terms corresponding to Fig. 6.2 gives:

$$S_2 = (iZ^{-1/2})^4(1/4\ i\chi_0)^2/2 \prod_{j=1}^{4} \int d^4y_j\, d^4y'_1\, d^4y'_2\, f_{p_1}(y_1)f_{p_2}(y_2)f_{p_3}^{\ *}(y_3)f_{p_4}^{\ *}(y_4)\cdot$$
$$\cdot(\Box_{y_1} + m^2)(\Box_{y_2}+ m^2)(\Box_{y_3}+ m^2)(\Box_{y_4}+ m^2)\cdot$$
$$\cdot[i\Delta_F^{\ TT}(y_1-y'_1)i\Delta_F^{\ TT}(y_2-y'_1)i\Delta_F^{\ TT}(y_3-y'_2)i\Delta_F^{\ TT}(y_4-y'_2\,) +$$
$$+ i\Delta_F^{\ TT}(y_1-y'_1)i\Delta_F^{\ TT}(y_2-y'_2)i\Delta_F^{\ TT}(y_3-y'_1)i\Delta_F^{\ TT}(y_4-y'_2\,)]\cdot$$
$$\cdot i\Delta_F^{\ TT}(y'_1-y'_2)i\Delta_F^{\ TT}(y'_1-y'_2\,) \tag{6.68}$$

Following a similar procedure to the previous calculation yields

$$S_2 = 4(iZ^{-1/2})^4(1/4\ i\chi_0)^2(2\pi)^4\delta^4(p_3 + p_4 - p_1 - p_2)\ \mathcal{N}^+(p_4)\mathcal{N}^+(p_3)\mathcal{N}(p_2)\mathcal{N}(p_1)\cdot$$
$$\cdot \int d^4z\, [e^{-i(p_1 + p_2).z} + e^{-i(p_1 - p_3).z}]\, [i\Delta_F^{\ TT}(z)]^2 \tag{6.69}$$

revealing a similar normalization of the external legs to that found in eq. 6.63, and a momentum conserving delta function as in eq. 6.63. The loop integrals have the form:

$$I(q) = \int d^4z\, e^{-iq.z}\, [i\Delta_F^{\ TT}(z)]^2 \tag{6.70}$$

The behavior of the two-tier Feynman propagator $\Delta_F^{\ TT}(z)$ was studied at long and short distance in eqs. 4.21-24. The large distance behavior of the two-tier Feynman propagator $\Delta_F^{\ TT}(z)$ approaches the behavior of the conventional Feynman propagator since

$$R(\mathbf{p},\, z) \to 1 \tag{6.71}$$

as $z^2 = z^\mu z_\mu$ becomes much larger than M_c^{-2} ($z^2 \gg M_c^{-2}$) in eq. 6.55. Thus $I(q)$ approaches the standard one loop expression of conventional field theory at large distance (or small momentum). *Two-tier quantum field theory realizes a form of Correspondence Principle approaching conventional quantum field theory at large distance.*

At short distances the Gaussian factor $R(\mathbf{p}, z)$ dominates. The two-tier Feynman propagator $\Delta_F{}^{TT}(z)$ is radically different from the conventional Feynman propagator at very short distances (or very high momentum). The singular behavior of the conventional Feynman propagator is replaced with a well-behaved, high-energy (short distance) behavior. Near the light cone $M_c^{-2} \gg z^2 \to 0$ (or $p^2 \gg M_c^2$) we can approximate eq. 6.52 with

$$i\Delta_F{}^{TT}(z) \approx \int d^3p\, [N(p)]^2\, R(\mathbf{p}, z) \tag{6.72}$$

since $e^{-ip\bullet z}$ is approximately unity for small z. We assume the mass of the ϕ particle is negligible on this scale. Upon performing the integrations (see eq. 4.23 for the exact result) we find eq. 6.72 approaches:

$$i\Delta_F{}^{TT}(z) \to \pi\, M_c^4 |z^2|/8 \tag{6.73}$$

as $z^2 = z^\mu z_\mu \to 0$ from the space-like or time-like side of the light cone where | | represents the absolute value.

Therefore $I(q)$ is finite and well-behaved. At high energy ($q^2 \gg M_c^2$)

$$I(q) \sim q^{-8} \tag{6.74}$$

since the fourier transform of $\Delta_F{}^{TT}(z)$ (momentum space)

$$\Delta_F{}^{TT}(p) = \int d^4z\, e^{-ip.z}\, \Delta_F{}^{TT}(z) \sim p^{-6} \tag{6.75}$$

for large p ($p^2 \gg M_c^2$). (Compare eq. 6.74 with the conventional logarithmically divergent one loop result $I(q) \sim \ln(q^2/\Lambda^2)$ with Λ a cutoff.)

Thus two-tier quantum provides the benefits of a higher derivative theory without its drawbacks.

General Form of Propagators

In this chapter we have considered a scalar two-tier quantum field theory. We have seen that the two-tier Feynman propagator is well behaved near the light cone resulting in a finite ϕ^4 theory. This finite ϕ^4 theory approximates the results of conventional ϕ^4 theory at low energy thus implementing a correspondence principle: *At low energy results in two-tier quantum field theory approach thecorresponding results of the corresponding conventional quantum field theory.*

The observations on two-tier field theory made in this chapter generally apply to two-tier versions of Quantum Electrodynamics, ElectroWeak Theory and the Standard Model as well as two-tier Quantum Gravity:

1. At low energy ($p^2 \ll M_c^2$ or large distances $z^2 \gg M_c^{-2}$) the two-tier quantum field theory is the same as the corresponding conventional quantum field theory to good approximation. (Correspondence Principle)
2. At high energy ($p^2 \gg M_c^2$ or short distances: $z^2 \ll M_c^{-2}$) two-tier quantum field theories (of physical interest) are well-behaved and finite.
3. Two-tier quantum field theories (of physical interest) are unitary and Lorentz invariant (and in the case of quantum gravity they satisfy the requirements of general relativity).

The generality of these results is based on:

1. The expansion of the S matrix in time ordered products of field operators.
2. Wick's Theorem
3. The general form of all particle propagators in two-tier quantum field theories. All particle Feynman propagators have the form:

$$iG_F^{TT}{}_{\dots}(y_1 - y_2) = \langle 0|T(\chi_{\dots}(X(y_1)),\chi_{\dots}(X(y_2)))|0\rangle \tag{6.76}$$

$$= \int d^4p\, iG_{F\dots}(p)e^{-ip\bullet(y_1 - y_2)}\, R(\mathbf{p}, y_1 - y_2) \tag{6.77}$$

where $iG_{F\dots}(p)$ is the conventional momentum space $\chi_{\dots}$ particle propagator, and where ... represents the relevant tensor and matrix indices. $R(\mathbf{p}, y_1 - y_2)$ introduces a damping factor in each particle propagator that eliminates divergences.

Scalar Particle Propagator

The two-tier propagator for the case of a free scalar particle is:

$$i\Delta_F^{TT}(y_1 - y_2) = \langle 0|T(\phi(X(y_1)),\phi(X(y_2)))|0\rangle \tag{6.51}$$

$$= i \int \frac{d^4p\, e^{-ip\bullet(y_1 - y_2)} R(\mathbf{p}, y_1 - y_2)}{(2\pi)^4 (p^2 - m^2 + i\varepsilon)} \qquad (6.52)$$

Since the mass m is not relevant at high energy we set m = 0. This enables us to obtain a more tractable expressions for the propagator. After some manipulation the scalar propagator can be represented as:

$$i\Delta_F{}^{TT}(z) = -\beta[16\pi^3(AB)^{½}]^{-1} \int_{-\infty}^{\infty} dy_1 \int_{-\infty}^{\infty} dy_2$$

$$\{\theta(z_0)\exp[-\beta((y_1 - z_0)^2B + (y_2 + z)^2A)/(4AB)] +$$

$$+ \theta(-z_0)\exp[-\beta((y_1 + z_0)^2B + (y_2 - z)^2A)/(4AB)]\}/(y_1{}^2 - y_2{}^2) \qquad (6.78)$$

with $\beta = 4\pi^2M_c{}^4\mathbf{z}^2$. Using

$$(y_1{}^2 - y_2{}^2)^{-1} = -0.5 \int_0^{\infty} dq_1 \int_{-\infty}^{\infty} dq_2\, \theta(q_1{}^2 - q_2{}^2)\, \exp[iq_1y_1 - iq_2y_2] \qquad (6.79)$$

we obtain the representation

$$i\Delta_F{}^{TT}(z^\mu) = (8\pi^2)^{-1} \int_0^{\infty} dq_1 \int_{-\infty}^{\infty} dq_2\, \theta(q_1{}^2 - q_2{}^2)$$

$$\exp\{\, iq_1|z_0| + iq_2z - [A'q_1{}^2 + B'q_2{}^2]/[\,\beta'(\mathbf{z}^2 - z_0{}^2)]\} \qquad (6.80)$$

where $|z_0|$ is the absolute value of z_0, $\mathbf{z}^2 - z_0{}^2 = -z^\mu z_\mu$ and

$$A = A'/(1 - v^2) \qquad (6.81)$$

$$B = B'/(1 - v^2) \qquad (6.82)$$

$$\beta = 4\pi^2M_c{}^4\mathbf{z}^2 = \beta'z^2 \qquad (6.83)$$

with $\mathbf{z} = |\vec{\mathbf{z}}|$ – the magnitude of the spatial vector $\vec{\mathbf{z}}$, and A and B given by eqs. 6.60 – 6.61.

The representation of $i\Delta_F{}^{TT}$ in eq. 6.80 is particularly useful in determining its low energy ($\ll M_c$), and its high energy ($\gg M_c$) behavior. The low energy behavior is governed by the linear terms in the exponential in eq. 6.80 since $\beta'(\mathbf{z}^2 - z_0{}^2)$ is very large in this limit:

$$i\Delta_F{}^{TT}(z^\mu)_{low} \simeq (8\pi^2)^{-1} \int_0^\infty dq_1 \int_{-\infty}^\infty dq_2\, \theta(q_1{}^2 - q_2{}^2)\exp\{ iq_1|z_0| + iq_2 z\} \qquad (6.84)$$

$$= [4\pi^2(\mathbf{z}^2 - z_0{}^2)]^{-1} \qquad (6.85)$$

$$= i\Delta_F(z^\mu)$$

equaling the exact massless, free, spin 0 Feynman propagator of conventional quantum field theory.

In the high energy limit when $\beta'(\mathbf{z}^2 - z_0{}^2)$ is small since $\mathbf{z}^2 \approx z_0{}^2$ (i.e. near the light cone), the quadratic terms in the exponential in eq. 6.80 dominate and $A' \simeq B'$. We then find

$$i\Delta_F{}^{TT}(z^\mu)_{high} \simeq (8\pi^2)^{-1} \int_0^\infty dq_1 \int_{-\infty}^\infty dq_2 \theta(q_1{}^2 - q_2{}^2)\exp\{ A'(q_1{}^2 + q_2{}^2)/[\beta'(\mathbf{z}^2 - z_0{}^2)]\} \qquad (6.86)$$

$$= \pi M_c{}^4|(\mathbf{z}^2 - z_0{}^2)|/8 \qquad (6.87)$$

as in eq. 4.24. As pointed out earlier, eq. 6.87 corresponds to k^{-6} behavior in momentum space:

$$i\Delta_F{}^{TT}(k)_{high} \backsim k^{-6} \qquad (6.87a)$$

Spin ½ Particle Propagator

For the case of a free, spin ½ particle the propagator is:

$$iS_F{}^{TT}(y_1 - y_2) = \langle 0|T(\bar{\psi}(X(y_1))\psi(X(y_2)))|0\rangle \qquad (6.88)$$

$$= i \int \frac{d^4p\, e^{-ip\bullet(y_1 - y_2)}\, (\not{p} + m)\, R(\mathbf{p}, y_1 - y_2)}{(2\pi)^4\, (p^2 - m^2 + i\varepsilon)}$$

Again setting m = 0 we find a convenient representation in the form:

$$S_F^{TT}(z^\mu) = i(8\pi^2)^{-1} \int_0^\infty dq_1 \int_{-\infty}^\infty dq_2\ \theta(q_1^2 - q_2^2)(\in(z_0)q_1\gamma_0 - q_2\vec{\mathbf{z}}\cdot\vec{\gamma}/z)\cdot$$

$$\exp\{ iq_1|z_0| + iq_2 z \ - [A'q_1^2 + B'q_2^2]/[\beta'(\mathbf{z}^2 - z_0^2)]\} \quad (6.89)$$

using the same symbols and notation as eq. 6.80, and with $\in(z_0) = +1$ if $z_0 \geq 0$ and -1 otherwise.

The representation of S_F^{TT} in eq. 6.89 is useful in determining its low energy ($\ll M_c$), and high energy ($\gg M_c$) behavior. The low energy behavior is governed by the linear terms in the exponential in eq. 6.89 since $\beta'(\mathbf{z}^2 - z_0^2)$ is large in this limit:

$$S_F^{TT}(z^\mu)_{low} \simeq (8\pi^2)^{-1} \int_0^\infty dq_1 \int_{-\infty}^\infty dq_2\ \theta(q_1^2 - q_2^2)(\in(z_0)q_1\gamma_0 - q_2\vec{\mathbf{z}}\cdot\vec{\gamma}/z)\cdot$$

$$\exp\{ iq_1|z_0| + iq_2 z\} \quad (6.90)$$

$$= \not{z}\, [2\pi^2(\mathbf{z}^2 - z_0^2)^2]^{-1} \quad (6.91)$$

$$= S_F(z^\mu)$$

equaling the exact massless, spin ½ Feynman propagator of conventional quantum field theory. If we had not set m = 0 initially, we would have obtained the usual massive, spin ½ Feynman propagator.

In the high energy limit when $\beta'(\mathbf{z}^2 - z_0^2)$ is small since $\mathbf{z}^2 \approx z_0^2$ (i.e. near the light cone), the quadratic terms in the exponential in eq. 6.89 dominate and $A' \simeq B'$. We then find

$$S_F^{TT}(z^\mu)_{high} \simeq (8\pi^2)^{-1} \int_0^\infty dq_1 \int_{-\infty}^\infty dq_2 \theta(q_1^2 - q_2^2)(\in(z_0)q_1\gamma_0 - q_2\vec{\mathbf{z}}\cdot\vec{\gamma}/z)$$

$$\exp\{ A'(q_1^2 + q_2^2)/[\beta'(\mathbf{z}^2 - z_0^2)]\} \quad (6.92)$$

$$= i(8\pi^2)^{-1}\{ \mathbf{z}^{-1}(\mathbf{z}^2 - z_0^2)^{3/2} 2^{3/2}\pi^{7/2} M_c^6 z_0\gamma_0 - - 4i(\mathbf{z}^2 - z_0^2)^2 \pi^5 M_c^8 \vec{\mathbf{z}}\cdot\vec{\gamma})\} \quad (6.93)$$

The leading momentum dependence of the fourier transform of $S_F^{TT}(z^\mu)_{high}$ is

$$S_F^{TT}(p)_{high} \backsim M_c^6 p^{-7} \gamma_0 \tag{6.94}$$

Massless Spin 1 Particle Propagator

The two-tier Feynman propagator for the case of a free, massless, spin 1, gauge field particle (coupled to a conserved current) such as a photon is:

$$iD_F^{TT}(z)_{\mu\nu} = -i \int \frac{d^4p\, e^{-ip\bullet z} g_{\mu\nu} R(\mathbf{p}, y_1 - y_2)}{(2\pi)^4 (p^2 + i\varepsilon)} \tag{6.95}$$

The form of eq. 6.95 is the same as the scalar particle propagator multiplied by $-g_{\mu\nu}$. As a result we have the representation:

$$iD_F^{TT}(z)_{\mu\nu} = -(8\pi^2)^{-1} \int_0^\infty dq_1 \int_{-\infty}^\infty dq_2\, \theta(q_1^2 - q_2^2)\, g_{\mu\nu}$$

$$\exp\{ iq_1|z_0| + iq_2 z - [A'q_1^2 + B'q_2^2]/[\beta'(\mathbf{z}^2 - z_0^2)]\} \tag{6.96}$$

As before in the scalar particle case, the low energy behavior is governed by the linear terms in the exponential in eq. 6.96 since $\beta'(\mathbf{z}^2 - z_0^2)$ is very large in this limit:

$$iD_F^{TT}(z)_{\mu\nu low} \simeq -g_{\mu\nu}(8\pi^2)^{-1} \int_0^\infty dq_1 \int_{-\infty}^\infty dq_2\, \theta(q_1^2 - q_2^2)\exp\{ iq_1|z_0| + iq_2 z\} \tag{6.97}$$

$$= -g_{\mu\nu}[4\pi^2(\mathbf{z}^2 - z_0^2)]^{-1} \tag{6.98}$$

$$= -ig_{\mu\nu}\Delta_F(z)$$

equaling the exact free, massless, spin 1 Feynman gauge field propagator of conventional quantum field theory.

In the high energy limit when $\beta'(\mathbf{z}^2 - z_0^2)$ is small since $\mathbf{z}^2 \approx z_0^2$ (i.e. near the light cone), the quadratic terms in the exponential in eq. 6.96 dominate, and $A' \simeq B'$. We then find

$$iD_F{}^{TT}(z)_{\mu\nu\text{high}} \simeq -(8\pi^2)^{-1}\int_0^{\infty} dq_1 \int_{-\infty}^{\infty} dq_2\,\theta(q_1{}^2-q_2{}^2)\, g_{\mu\nu}$$

$$\exp\{\, A'(q_1{}^2 + q_2{}^2)/[\,\beta'(\mathbf{z}^2 - z_0{}^2)]\} \qquad (6.99)$$

$$= -g_{\mu\nu}\pi\, M_c{}^4\, |(\mathbf{z}^2 - z_0{}^2)|/8 \qquad (6.100)$$

Eq. 6.100 corresponds to k^{-6} behavior in momentum space:

$$iD_F{}^{TT}(k)_{\mu\nu\text{high}} \backsim g_{\mu\nu}\, M_c{}^4 k^{-6} \qquad (6.101)$$

Spin 2 Particle Propagator

The two-tier propagator for the case of a free, massless, spin 2 particle such as a graviton is:

$$i\Delta_{F2}{}^{TT}(z)_{\mu\nu\rho\sigma} = i\int \frac{d^4p\, e^{-ip\cdot z}\, b_{\mu\nu\rho\sigma}(p) R(\mathbf{p}, y_1 - y_2)}{(2\pi)^4\,(p^2 + i\varepsilon)} \qquad (6.102)$$

in an appropriate gauge where $b_{\mu\nu\rho\sigma}(p)$ is a tensor that is independent of the coordinates. We can express eq. 6.102 in the form:

$$i\Delta_{F2}{}^{TT}(z^\mu)_{\mu\nu\rho\sigma} = (8\pi^2)^{-1}\int_0^{\infty} dq_1 \int_{-\infty}^{\infty} dq_2\; \theta(q_1{}^2 - q_2{}^2)\, \tilde{b}(z_0, \mathbf{z}, q_1, q_2)_{\mu\nu\rho\sigma}$$

$$\exp\{\, iq_1|z_0| + iq_2 z - [A'q_1{}^2 + B'q_2{}^2]/[\,\beta'(\mathbf{z}^2 - z_0{}^2)]\} \quad (6.103)$$

where $\tilde{b}(z_0, \mathbf{z}, q_1, q_2)_{\mu\nu\rho\sigma}$ is a tensor generated from the $b_{\mu\nu\rho\sigma}(p)$ tensor.

As before in the scalar particle case, the low energy behavior is governed by the linear terms in the exponential in eq. 6.103 since $\beta'(\mathbf{z}^2 - z_0{}^2)$ is very large in this limit and we find that the covariant piece[25] behaves like:

[25] S. Weinberg, Phys. Rev. **135**, B1049 (1964); Phys. Rev. **138**, B988 (1965).

$$i\Delta_{F2}{}^{TT}(z^{\mu})_{\mu\nu\rho\sigma\text{lowCov}} \simeq \tilde{\tilde{b}}_{\mu\nu\rho\sigma}(8\pi^2)^{-1}\int_0^{\infty} dq_1 \int_{-\infty}^{\infty} dq_2\ \theta(q_1{}^2 - q_2{}^2)\exp\{\ iq_1|z_0| + iq_2 z\} \tag{6.104}$$

$$= \tilde{\tilde{b}}_{\mu\nu\rho\sigma}[4\pi^2(\mathbf{z}^2 - z_0{}^2)]^{-1} \tag{6.105}$$

$$= i\Delta_F(z^{\mu})\ \tilde{\tilde{b}}_{\mu\nu\rho\sigma}$$

where

$$\tilde{\tilde{b}}_{\mu\nu\rho\sigma} = ½\ [\eta_{\mu\rho}\eta_{\nu\sigma} + \eta_{\mu\sigma}\eta_{\nu\rho} - \eta_{\mu\nu}\eta_{\rho\sigma}] \tag{6.106}$$

so that the expression in eq. 6.105 equals the corresponding covariant piece of the exact free, massless, spin 2 Feynman propagator of conventional quantum field theory.

In the high energy limit when $\beta'(\mathbf{z}^2 - z_0{}^2)$ is small since $\mathbf{z}^2 \approx z_0{}^2$ (i.e. near the light cone), the quadratic terms in the exponential in eq. 6.103 dominate, and $A' \simeq B'$. We then find

$$i\Delta_{F2}{}^{TT}(z^{\mu})_{\mu\nu\rho\sigma\text{high}} \simeq (8\pi^2)^{-1}\int_0^{\infty} dq_1 \int_{-\infty}^{\infty} dq_2\theta(q_1{}^2-q_2{}^2)\ \tilde{b}(z_0,\ z,\ q_1,\ q_2)_{\mu\nu\rho\sigma}$$

$$\exp\{\ A'(q_1{}^2 + q_2{}^2)/[\ \beta'(\mathbf{z}^2-z_0{}^2)]\} \tag{6.107}$$

and the covariant piece behaves like

$$i\Delta_{F2}{}^{TT}(z^{\mu})_{\mu\nu\rho\sigma\text{highCov}} \simeq \tilde{\tilde{b}}_{\mu\nu\rho\sigma}\pi M_c{}^4|(\mathbf{z}^2 - z_0{}^2)|/8 \tag{6.108}$$

The coordinate space behavior of eq. 6.108 corresponds to k^{-6} behavior in momentum space:

$$i\Delta_{F2}{}^{TT}(k)_{\mu\nu\rho\sigma\text{highCov}} \backsim \tilde{\tilde{b}}_{\mu\nu\rho\sigma}\ k^{-6} \tag{6.109}$$

The high-energy behavior of the spin 2 propagator in momentum space results in a two-tier theory of quantum gravity that has no high-energy divergences and is thus finite. See chapter 9 for a detailed discussion.

7. Two-Tier Quantum Electrodynamics

Formulation

There have been numerous attempts to develop a finite theory of Quantum Electrodynamics (QED). Among the noteworthy attempts are the Lee-Wick[26] formulation of QED and the Johnson-Baker-Willey model.[27] The unification of the electromagnetic interaction with the weak interaction in the Electroweak Theory and the proof that it is renormalizable, has switched the focus of interest away from QED. However the extremely precise experimental tests of QED, which are among the most accurate measurements made by science, and the impressive agreement[28] with the theoretical predictions of QED, make QED of interest in its own right.

In this chapter we will describe the formulation of two-tier QED. We will see that it is finite, and yet it is in complete agreement with the highly accurate calculations of QED if M_c is sufficiently large – such as of the order of the Planck mass. We will also see a modification of the Coulomb potential in the two-tier model that makes possible the existence of (unstable) bound states of particles with the same charge such as a two electron bound state. The two-tier QED Coulomb potential is linear at short distances and zero at r =0.

A New Quantum Electrodynamics with Non-Commuting Coordinates

Two-tier QED is formulated in a way that is similar to conventional QED in order to capture the excellent results of QED while making the theory finite. We will consider the case of QED for electrons. The results apply directly to any charged spin ½ field and with a few trivial changes to charged particles of differing spin. The two-tier QED Lagrangian that we will investigate is:

---

[26] T. D. Lee and G. C. Wick, Phys. Rev. **D2**, 1033 (1970); T. D. Lee and G. C. Wick, Nucl. Phys. **B9**, 209 (1969) and references therein.

[27] S. Blaha, "An Approximate Calculation of the Eigenvalue Function in Massless Quantum Electrodynamics", Phys.Rev. **D9**, 2246 (1974) and references therein.

[28] T. Kinoshita, "The Fine Structure Constant", Cornell University preprint CLNS 96/1406 (1996); V. W. Hughes and T. Kinoshita, Rev. Mod. Phys. **71**, S133 (1999).

$$\mathscr{L} = J\,\mathscr{L}_F + \mathscr{L}_C(X^\mu(y), \partial X^\mu(y)/\partial y^\nu, y) \tag{7.1}$$

with J the Jacobian and

$$\mathscr{L}_F = \bar{\psi}(X(y))(i\not{\nabla}_X - e_0\not{A}(X(y)) - m_0)\psi(X(y)) - \tfrac{1}{4}F^{\mu\nu}(X(y))F_{\mu\nu}(X(y)) \tag{7.2}$$

with

$$F_{\mu\nu}(X(y)) = \partial A_\mu(X(y))/\partial X^\nu - \partial A_\nu(X(y))/\partial X^\mu \tag{7.3}$$

and

$$\mathscr{L}_C(X^\mu(y), \partial X^\mu(y)/\partial y^\nu, y) = -\tfrac{1}{4}F_Y{}^{\mu\nu}F_{Y\mu\nu} \tag{7.4}$$

$$F_{Y\mu\nu} = \partial Y_\mu/\partial y^\nu - \partial Y_\nu/\partial y^\mu \tag{7.5}$$

We note the Lagrangian $\mathscr{L}_F$ has the form of the conventional electromagnetic Lagrangian[29] except for the functional dependence on X(y).

Since the Lagrangian in eq. 7.2 is separable we will follow the same procedure as we did for the scaiar field theory in the development of eqs. 2.74 – 2.112. Thus we obtain the hamiltonian (as in eq. 2.112 for the scalar case):

$$H_F = \int d^3X\,(\mathscr{H}_{F0} + \mathscr{H}_{Fint}) \tag{7.6}$$

where

$$\mathscr{H}_{F0} = \bar{\psi}(X(y))(i\not{\nabla}_X - m_0)\psi(X(y)) + \tfrac{1}{2}(\mathbf{E}^2 + \mathbf{B}^2) \tag{7.7}$$

$$\mathscr{H}_{Fint} = e_0\bar{\psi}(X(y))\not{A}(X(y))\psi(X(y)) \tag{7.8}$$

with **E** being the electric field and **B** being the magnetic field:

$$E^i = -\partial A^i/\partial y^0$$

---

[29] We follow the conventions of Kaku, and Bjorken and Drell. It will be evident that the proof that two-tier QED is finite will not require detailed knowledge of specific conventions.

$$B^i = \varepsilon^{ijk}\, \partial A_j/\partial y^k$$

The field equations are:

$$(i\not\nabla_X - e_0 \not A(X(y)) - m_0)\psi(X(y)) = 0 \tag{7.9}$$

and

$$\partial(F^{\mu\nu}(X(y)))/\partial X^\nu = e_0\bar{\psi}(X(y))\gamma^\mu\psi(X(y)) \tag{7.10}$$

The Y Field

The X^μ coordinate field is related to the Y^μ field via:

$$X_\mu(y) = y_\mu + i\, Y_\mu(y)/M_c^2 \tag{3.12}$$

The Y field has the free Lagrangian eq. 7.4 (based on eqs. 3.10 – 3.14), and the hamiltonian

$$\mathscr{H}_C = ½\, (E_Y^2 + B_Y^2) \tag{7.11}$$

where

$$E_Y^i = -\partial Y^i/\partial y^0 \tag{7.12}$$

$$B_Y^i = \varepsilon^{ijk}\, \partial Y_j/\partial y^k \tag{7.13}$$

The Y field equation is

$$\partial F_Y^{\mu\nu}(y)/\partial y^\nu = 0 \tag{7.14}$$

The quantization of the Y field in the Coulomb gauge is described in chapter 3. We will use the Y Coulomb gauge throughout our discussions of various two-tier quantum field theories.

Quantization of the Free Dirac Field

The quantization procedure is formally identical to that of conventional QED. The standard equal time anti-commutation relations for the spin ½ field are:

$$\{\psi_\alpha(X), \psi_\beta(X')\} = \{\pi_{\psi\alpha}(X), \pi_{\psi\beta}(X')\} = 0 \tag{7.15}$$

$$\{\pi_{\psi\alpha}(X), \psi_\beta(X')\} = i\,\delta_{\alpha\beta}\,\delta^3(\mathbf{X} - \mathbf{X}') \tag{7.16}$$

where α and β are the spinor indices and where

$$\pi_{\psi\alpha}(X) = i\,\psi_\alpha^{\dagger}(X) \tag{7.17}$$

The spin ½ field can be expanded in a fourier series:

$$\psi(X(y)) = \sum_{\pm s} \int d^3p\, N^d{}_m(p)\, [b(p,s)u(p,s)\, :e^{-ip\cdot(y + iY/M_c{}^2)}: + \\ + d^{\dagger}(p,s)v(p,s)\, :e^{ip\cdot(y + iY/M_c{}^2)}:] \tag{7.18}$$

$$\psi^{\dagger}(X(y)) = \sum_{\pm s} \int d^3p\, N^d{}_m(p)\, [b^{\dagger}(p,s)\bar{u}(p,s)\gamma^0\, :e^{+ip\cdot(y + iY/M_c{}^2)}: + \\ + d(p,s)\bar{v}(p,s)\gamma^0\, :e^{-ip\cdot(y + iY/M_c{}^2)}:] \tag{7.19}$$

where

$$N^d{}_m(p) = [m/((2\pi)^3E_p)]^{½} \tag{7.20}$$

and

$$E_p = (\mathbf{p}^2 + m^2)^{½} \tag{7.21}$$

The commutation relations of the Fourier coefficient operators are:

$$\{b(p,s),\, b^{\dagger}(p',s')\} = \delta_{ss'}\,\delta^3(\mathbf{p} - \mathbf{p}') \tag{7.22}$$

$$\{d(p,s),\, d^{\dagger}(p',s')\} = \delta_{ss'}\,\delta^3(\mathbf{p} - \mathbf{p}') \tag{7.23}$$

$$\{b(p,s),\, b(p',s')\} = \{d(p,s),\, d(p',s')\} = 0 \tag{7.24}$$

$$\{b^{\dagger}(p,s),\, b^{\dagger}(p',s')\} = \{d^{\dagger}(p,s),\, d^{\dagger}(p',s')\} = 0 \tag{7.25}$$

$$\{b(p,s),\, d^{\dagger}(p',s')\} = \{d(p,s),\, b^{\dagger}(p',s')\} = 0 \tag{7.26}$$

$$\{b^{\dagger}(p,s),\, d^{\dagger}(p',s')\} = \{d(p,s),\, b(p',s')\} = 0 \tag{7.27}$$

The spinors u(p,s) and v(p,s) are defined in the conventional way (as in Kaku, and in Bjorken and Drell).

Quantization of the Electromagnetic Field

The gauge invariance of two-tier QED can be seen by examining the field equation

$$(i\not{\partial}_X - e_0\not{A}(X(y)) - m_0)\psi(X(y)) = 0 \qquad (7.9)$$

and considering the effect of a gauge gauge transformation:

$$A^\mu(X(y)) \rightarrow A^\mu(X(y)) - \partial\Lambda(X(y))/\partial X_\mu \qquad (7.28)$$

The field equation eq. 7.10 remains unchanged and the change in eq. 7.9 can be accommodated by a change of phase of the Dirac field:

$$\psi(X(y)) \rightarrow \exp(ie_0\Lambda(X(y))\ \psi(X(y)) \qquad (7.29)$$

The only novelty is that $\Lambda(X(y))$ in general becomes a complex q-number quantity at extremely short distances since X is a complex q-number.

The gauge invariance of the Lagrangian eq. 7.2 allows us to choose a convenient gauge. It appears that the most convenient gauge is the Coulomb gauge[30]:

$$\partial A^i/\partial X^i = 0 \qquad (7.30)$$

where the sum is over spatial components labeled with i. We also set

$$A^0 = 0 \qquad (7.31)$$

in the absence of an external source.

A conventional treatment leads to the equal time commutation relations:

$$[A^\mu(X(y)), A^\nu(X(y'))] = [\pi_A{}^\mu(X(y)), \pi_A{}^\nu(X(y'))] = 0 \qquad (7.32)$$

[30] It is also possible to quantize in two-tier QED using an indefinite metric that preserves manifest Lorentz covariance as was done by Gupta and Bleuler for conventional QED.

$$[\pi_A{}^j(X(y)),\ A_k(X(y'))] = -i\,\delta^{tr}{}_{jk}(\mathbf{X}(y) - \mathbf{X}(y')) \tag{7.33}$$

(note the locations of the j component label in eq. 7.33 introduces a minus sign) where

$$\pi_A{}^k = \partial\mathscr{L}_F/\partial\dot{A}_k \tag{7.34}$$

$$\pi_A{}^0 = 0 \tag{7.35}$$

$$\delta^{tr}{}_{jk}(X(y) - X(y')) = \int d^3k\ e^{i\,\mathbf{k}\bullet(\mathbf{X}(y) - \mathbf{X}(y'))}\,(\delta_{jk} - k_j k_k/\mathbf{k}^2)/(2\pi)^3 \tag{7.36}$$

$$\dot{A}_k = \partial A_k/\partial X^0 \tag{7.37}$$

The Coulomb gauge reveals the two transverse degrees of freedom that are present in the vector potential. The Fourier expansion of the vector potential is:

$$A^i(X(y)) = \int d^3k\ N_0(k) \sum_{\lambda=1}^{2} \varepsilon^i(k, \lambda)[a(k,\lambda)\ e^{-ik.X(y)} + a^\dagger(k,\lambda)\ e^{ik.X(y)}] \tag{7.38}$$

where

$$N_0(k) = [(2\pi)^3 2\omega_k]^{-1/2} \tag{7.39}$$

(m = 0) and

$$\omega_k = (\mathbf{k}^2)^{1/2} = k^0 \tag{7.40}$$

with $\epsilon(k, \lambda)$ being the polarization unit vectors for $\lambda = 1, 2$.

The commutation relations of the Fourier coefficient operators are:

$$[a(k,\lambda),\ a^\dagger(k',\lambda')] = \delta_{\lambda\lambda'}\,\delta^3(\mathbf{k} - \mathbf{k}') \tag{7.41}$$

$$[a^\dagger(k,\lambda),\ a^\dagger(k',\lambda')] = [a(k,\lambda),\ a(k',\lambda')] = 0 \tag{7.42}$$

and the polarization vectors satisfy

$$\sum_{\lambda=1}^{2} \varepsilon_i(k, \lambda)\varepsilon_j(k, \lambda) = (\delta_{ij} - k_i k_j/\mathbf{k}^2) \tag{7.43}$$

Particle propagators

The electron and photon Feynman propagators differ from the conventional QED propagators by having the Gaussian factor $\mathrm{R}(\mathbf{p}, \mathbf{z})$ in their fourier expansions:

$$iS_F^{TT}(y_1 - y_2) = <0|T(\bar{\psi}(X(y_1))\psi(X(y_2)))|0> \tag{7.44}$$

where the time ordering is with respect to y_1^0 and y_2^0. Expanding the free fields leads to the fourier representation:

$$iS_F^{TT}(y_1 - y_2) = i \int \frac{d^4p\, e^{-ip\bullet(y_1 - y_2)}\, (\not{p} + m)\, R(\mathbf{p}, y_1 - y_2)}{(2\pi)^4\, (p^2 - m^2 + i\varepsilon)} \tag{7.45}$$

with the Gaussian factor $\mathrm{R}(\mathbf{p}, \mathbf{z})$ specified in eq. 6.53. The photon propagator is

$$iD_F^{trTT}(y_1 - y_2)_{\mu\nu} = <0|T(A_\mu(X(y_1))A_\nu(X(y_2)))|0> \tag{7.46}$$

$$= -\, ig_{\mu\nu} \int \frac{d^4k\, e^{-ik\bullet(y_1 - y_2)}\, R(\mathbf{k}, y_1 - y_2)}{(2\pi)^4\, (k^2 + i\varepsilon)} \tag{7.47}$$

plus gauge terms and minus the Coulomb term. The presence of the Gaussian factor $\mathrm{R}(\mathbf{p}, \mathbf{z})$ results in a theory of QED that has no divergences and thus is finite. See eqns. 6.94 and 6.101 for their large momentum behavior.

Coulomb Interaction

The Coulomb interaction in two-tier QED is different at short distances from the conventional Coulomb interaction. The Coulomb potential between singly charged (same sign) particles in two-tier QED is:

$$V_{TT}\,(y_1 - y_2) = e^2 \int \frac{d^4p\, e^{-ip\bullet(y_1 - y_2)}\, R(\mathbf{p}, y_1 - y_2)}{(2\pi)^4\, \mathbf{k}^2} \tag{7.48}$$

$$= \alpha\, \Phi(M_c^2 \pi r^2)\, \delta(y_1^0 - y_2^0)/r \qquad (7.49)$$

where α is the fine structure constant, $\Phi(z)$ is the error function, M_c is the mass setting the scale of the new short distance behavior and $r = \sqrt{(\mathbf{y_1} - \mathbf{y_2})^2}$ is the radial distance. At small distances ($\pi r^2 \ll M_c^{-2}$) the two-tier potential becomes linear in r:

$$V_{TT} \to \alpha 2\sqrt{\pi}\, M_c^2\, r\, \delta(y_1^0 - y_2^0) \qquad (7.50)$$

and at large distances ($\pi r^2 \gg M_c^{-2}$) the two-tier potential approaches the conventional Coulomb potential:

$$V_{TT} \to V_{Coul} = \alpha\, \delta(y_1^0 - y_2^0)/r \qquad (7.51)$$

using the error function normalization $\Phi(\infty) = 1$. The modified Coulomb potential V_{TT} of eq. 7.49 is plotted in Fig. 1.1 using $M_c = 1000$ GeV/c^2. At large distances the Coulomb potential (which has been verified experimentally with great precision) can be approximated arbitrarily closely by two-tier QED by simply letting M_c become larger. Conceivably M_c can be as large as the Planck mass (1.221×10^{19} GeV/c^2) or even larger. Thus conventional QED is the large distance limit of two-tier QED.

The short distance behavior of the two-tier Coulomb potential opens the possibility of quasi-bound states of particles of the same sign such as a two electron bound state. The normally repulsive potential has a linear behavior near r = 0 and a potential barrier before becoming like the conventional Coulomb potential at larger distances. A pair of electrons, if localized within the linear region of the potential, would be bound but the bound state would quickly decay via electron tunneling through the barrier. States of this type conceivably might have existed in the first instants after the Big Bang and influenced the earliest evolution of the universe. Creating these dilepton states does not appear to be feasible if M_c is extremely large.

Asymptotic States

The development of perturbation theory in chapter 6 applies to the two-tier theory of QED with only superficial changes.

First we note that the form of the photon propagator has exactly the form of eq. 7.47 since terms proportional to k_μ or k_ν that would appear in the evaluation of eq. 7.46 do not contribute due to current conservation. In addition the instantaneous Coulomb interaction cancels the remaining Coulomb-like term appearing in the evaluation of eq. 7.46.

Thus we are left with the electron propagator (eq. 7.45) and the effective photon propagator (eq. 7.47). The formalism and role of the Y field is the same as in the scalar two-tier quantum field theory considered earlier.

In-states and Out-states

The dependence of the Lagrangian, and the particle fields in particular, on X^μ rather than directly on the coordinates y^μ leads to a "fuzziness" of the definition of asymptotic particle states that we have chosen to resolve with the construction of an asymptotic free field for each particle species of the "normal" sort. In the scalar case we defined an auxiliary field $\Phi_{in}(y)$ using the raising and lowering operators of the free scalar field $\phi_{in}(X(y))$. In actuality $\Phi_{in}(y) \equiv \phi_{in}(y)$. The change from the argument y^μ to $X^\mu(y)$ is a form of translation that can be implemented using the (non-unitary) exponentiated momentum operator as we did in eq. 5.70:

$$\phi_a(y) \equiv \Phi_a(y) = W_a^{-1}(y)\phi_a(X(y))W_a(y)$$

for a = "in" or "out." *The benefit from this approach is a clean simple definition of asymptotic particle states of definite momentum (and spin etc.).*

We will follow the same strategy in two-tier QED but we will not introduce auxiliary fields since we realize now that they are not necessary.

Fermion In-states and Out-States

In this section we will develop properties of fermion in-fields and out-fields. The LSZ procedure can be schematized as:

$$\psi_{in}(y) \Rightarrow \psi_{in}(X(y)) \Rightarrow \psi(X(y)) \Rightarrow \psi_{out}(X(y)) \Rightarrow \psi_{out}(y) \qquad (7.52)$$

In-states are constructed using $\psi_{in}(y)$ which is then transformed into $\psi_{in}(X(y))$ in order to make contact with our Lagrangian formalism. The interacting field $\psi(X(y))$ is related to $\psi_{in}(X(y))$ using the standard LSZ limiting ($y^0 \to -\infty$) process. Similarly out-states are constructed using $\psi_{out}(y)$ which is transformed into $\psi_{out}(X(y))$. Again the interacting field $\psi(X(y))$ is related to $\psi_{out}(X(y))$ as part of the familiar LSZ limiting ($y^0 \to +\infty$) process.

Since much of the development differs only trivially from the standard treatment in textbooks we will simply list relevant equations and let the reader pursue them further in quantum field theory textbooks.

ψ In-Field

In order to define a perturbation theory for particle scattering we will use a free Dirac field $\psi_{in}(y)$ that satisfies

$$(i\not{\partial}_y - m)\psi_{in}(y) = 0 \qquad (7.53)$$

where

$$\not{\nabla}_y = \gamma^\nu \partial/\partial y^\nu$$

Defining the fourier expansion

$$\psi_{in}(y) = \sum_{\pm s} \int d^3p\ N^d{}_m(p)\ [b_{in}(p,s)u(p,s)\ e^{-ip.y} + d_{in}{}^\dagger(p,s)v(p,s)\ e^{ip.y}] \quad (7.54)$$

we can define ψ_{in} in-field states with expressions like

$$|(p_n s_n),\ \ldots,\ (p_1 s_1);\ (\bar{p}_m \bar{s}_m),\ \ldots\ (\bar{p}_1 \bar{s}_1)\ in> = b_{in}{}^\dagger(p_n s_n)\ \ldots\ b_{in}{}^\dagger(p_1 s_1)\ d_{in}{}^\dagger(\bar{p}_m \bar{s}_m)\ \ldots\ d_{in}{}^\dagger(\bar{p}_1 \bar{s}_1)|0> \quad (7.55)$$

for n electrons and m positrons. The development parallels the conventional development of fermion in-states.

ψ Out-Field

Similarly we can define fermion out-states for the free field $\psi_{out}(y)$. Again we will use a free Dirac field $\psi_{out}(y)$ that satisfies

$$(i\not{\nabla}_y - m)\psi_{out}(y) = 0 \quad (7.56)$$

Defining the fourier expansion

$$\psi_{out}(y) = \sum_{\pm s} \int d^3p\ N^d{}_m(p)\ [b_{out}(p,s)u(p,s)\ e^{-ip.y} + d_{out}{}^\dagger(p,s)v(p,s)\ e^{ip.y}] \quad (7.57)$$

We can define ψ_{out} out-field states with expressions like

$$|(p_n s_n),\ \ldots,\ (p_1 s_1);\ (\bar{p}_m \bar{s}_m),\ \ldots\ (\bar{p}_1 \bar{s}_1)\ out> = b_{out}{}^\dagger(p_n s_n)\ \ldots\ b_{out}{}^\dagger(p_1 s_1)\ d_{out}{}^\dagger(\bar{p}_m \bar{s}_m)\ \ldots\ d_{out}{}^\dagger(\bar{p}_1 \bar{s}_1)|0> \quad (7.58)$$

for n electrons and m positrons. The development parallels the conventional development of fermion out-states.

Photon In and Out States

In this section we will develop properties of photon in-fields and out-fields. The LSZ procedure can be schematized as:

$$A_{in}^{\mu}(y) \Rightarrow A_{in}^{\mu}(X(y)) \Rightarrow A^{\mu}(X(y)) \Rightarrow A_{out}^{\mu}(X(y)) \Rightarrow A_{out}^{\mu}(y) \qquad (7.59)$$

In-states are constructed using $A_{in}^{\mu}(y)$ which are then transformed into $A_{in}^{\mu}(X(y))$ in order to make contact with our Lagrangian formalism. The interacting field $A^{\mu}(X(y))$ is related to $A_{in}^{\mu}(X(y))$ using the standard LSZ limiting ($y^0 \rightarrow -\infty$) process. Similarly out-states are constructed using $A_{out}^{\mu}(y)$ which are transformed into $A_{out}^{\mu}(X(y))$. Again the interacting field $A^{\mu}(X(y))$ is related to $A_{out}^{\mu}(X(y))$ as part of the familiar LSZ limiting ($y^0 \rightarrow +\infty$) process.

The fourier expansion of the free photon in and out fields are

$$A_a^{i}(y) = \int d^3k\, N_0(k) \sum_{\lambda=1}^{2} \varepsilon^{i}(k, \lambda)[a_a(k,\lambda)\, e^{-ik\cdot y} + a_a^{\dagger}(k,\lambda)\, e^{ik\cdot y}] \qquad (7.60a)$$

$$A_a^{i}(X(y)) = \int d^3k\, N_0(k) \sum_{\lambda=1}^{2} \varepsilon^{i}(k, \lambda)[a_a(k,\lambda)\, e^{-ik\cdot X(y)} + a_a^{\dagger}(k,\lambda)\, e^{ik\cdot X(y)}] \qquad (7.60b)$$

where a = “in” or ‘out.”

We can define photon in-field states with expressions like

$$|(k_n\lambda_n), \ldots, (k_1\lambda_1)\ in\rangle = a_{in}^{\dagger}(k_n\lambda_n) \ldots a_{in}^{\dagger}(k_1\lambda_1)|0\rangle \qquad (7.61)$$

for n photons. The development parallels the conventional development of photon in-states as does the definition of photon out-states.

S Matrix

The S matrix is defined in a familiar way by

$$\psi_{in}(y) = S\psi_{out}(y)S^{-1} \qquad (7.62)$$

$$A^{\mu}_{in}(y) = SA^{\mu}_{out}(y)S^{-1} \qquad (7.63)$$

and the other standard properties of the S matrix with the sole exception being the form of the unitarity relation (discussed in the previous chapter).

LSZ Reduction

In this section we will determine the reduction formula for fermions and photons for the S matrix in two-tier QED.

Dirac Fields

Consider a charged Dirac particle such as an electron. The S matrix element corresponding to an in-state: β plus one Dirac particle of momentum p and spin s, and an out state α can be represented by

$$S_{\alpha\beta ps} = <\alpha \text{ out}|\beta \text{ (ps) in}> \qquad (7.64)$$

which becomes

$$S_{\alpha\beta ps} = <\alpha - \text{(ps) out}|\beta \text{ in}> + <\alpha \text{ out}|\int d^3y\, U_{ps}(y)[\psi_{in}^{\dagger}(y) - \psi_{out}^{\dagger}(y)]|\beta \text{ in}> \qquad (7.65)$$

through standard manipulations where $<\alpha - \text{(ps) out}|$ is an out state with a particle of momentum p and spin s removed (if present) and where

$$U_{ps}(y) = \{m/[(2\pi)^3E_p]\}^{1/2}u(p,s)e^{-ip.y} \qquad (7.66a)$$

and for later use

$$V_{ps}(y) = \{m/[(2\pi)^3E_p]\}^{1/2}v(p,s)e^{ip.y} \qquad (7.66b)$$

We now express

$$S_{\alpha\beta ps}=S_{\alpha-ps\beta}+<\alpha \text{ out}|\int d^3yU_{ps}(y)W_{QED}^{-1}[\psi_{in}^{\dagger}(X(y))-\psi_{out}^{\dagger}(X(y))]W_{QED}|\beta \text{ in}> \qquad (7.67)$$

using $W_{QED}(y) = W_{QEDin}(y)$ with

$$\psi_a(y) = W_{QEDa}^{-1}(y)\psi_a(X(y))W_{QEDa}(y) \qquad (7.68a)$$

$$\psi_a^{\dagger}(y) = VW_{QEDa}^{-1}(y)\psi_a^{\dagger}(X(y))W_{QEDa}(y)V \qquad (7.68b)$$

where the label a = "in" or a = "out", and where

$$W_{QEDa}(y) = \exp(\mathbf{Y}(y)\cdot\mathbf{P}_{QEDa}/M_c^2) \qquad (7.69)$$

and

$$W_{QEDa}^{-1}(y) = \exp(-\mathbf{Y}(y)\cdot\mathbf{P}_{QEDa}/M_c^2) \qquad (7.70)$$

in the Coulomb gauge of Y with $\mathbf{P}_{QEDa}$ being the spatial momentum vector for the free fermion and photon fields defined by

$$\mathbf{P}_{QEDa} = \sum_{\pm s} \int d^3p\, \mathbf{p}\, [b_a^\dagger(p,s)b_a(p,s) + d_a^\dagger(p,s)d_a(p,s)] + \sum_{\lambda=1}^{2} \int d^3k\, \mathbf{k}\, a_a^\dagger(k,\lambda)a_a(k,\lambda) \qquad (7.71)$$

using the free fermion and photon raising and lowering operators.

We note that $W_{QEDa}(y)$ is not a unitary operator (A similar situation to that of the scalar particle quantum field theory.) but it is pseudo-unitary:

$$W_{QEDa}^{-1}(y) = V\, W_{QEDa}^\dagger(y)\, V^{-1} \qquad (7.72)$$

where (letting $a_Y^\dagger(k, \lambda)$ and $a_Y(k, \lambda)$ represent the raising and lowering operators of the Y field)

$$V = \exp(-i\pi \sum_{\lambda=1}^{2} \int d^3k\, a_Y^\dagger(k, \lambda)a_Y(k, \lambda)) \qquad (7.73)$$

is a unitary operator with the property

$$V\, Y^j(y)\, V^{-1} = -Y^j(y) \qquad (7.74)$$

for j = 1,2,3. We note (as in the scalar field discussion)

$$V^\dagger = V^{-1} = V \qquad (7.75)$$

and thus

$$V^2 = I \qquad (7.76)$$

V is a metric operator in the sense of Dirac as discussed earlier. We also note:

$$X^\mu(y) = V[X^\mu(y)]^\dagger V^{-1} \qquad (7.77)$$

$$\psi_a^\dagger(X(y)) = V[\psi_a(X(y))]^\dagger V^{-1} \qquad (7.78)$$

$$A^{\mu}{}_{a}(X(y)) = V[A^{\mu}{}_{a}(X(y))]^{\dagger} V^{-1} \qquad (7.79)$$

for a = "in" or "out." These properties are required for eqs. 7.68a and 7.68b to hold.

The interacting $\psi(X(y))$ approaches the in and out fields $\psi_{in}(X(y))$ and $\psi_{out}(X(y))$ in the limit that $y^0 \to -\infty$ and $y^0 \to +\infty$ respectively in the sense of Lehmann, Symanzik and Zimmermann which we *symbolize* as:

$$\psi(X(y)) \to \sqrt{Z_2}\, \psi_{in}(X(y)) \quad \text{as} \quad y^0 \to -\infty \qquad (7.72)$$

$$\psi(X(y)) \to \sqrt{Z_2}\, \psi_{out}(X(y)) \quad \text{as} \quad y^0 \to +\infty \qquad (7.73)$$

with $\sqrt{Z_2}$ the wave function renormalization constant. Using eqs. 7.72 and 7.73 and following the standard LSZ reduction procedure leads to:

$$S_{\alpha\beta ps} = S_{\alpha-ps\beta} - iZ_2^{-1/2}\int d^4y <\alpha\ out|W_{QED}^{-1}\bar{\psi}(X(y))W_{QED}|\beta\ in>(-i\overleftarrow{\not\nabla}_y - m)U_{ps}(y)$$
$$(7.74)$$

Eq. 7.74 is similar to the usual LSZ reduction formula for a fermion extracted from an in-state except for the appearance of the W(y) operator and its inverse. We note that W(y) = $W_{in}(y)$ still because $\mathbf{P}_{QEDin}$ is independent of y^0.

The expressions for the other possible reductions of a fermion and its anti-particle are:

1. Reduction of an anti-particle from an in-state

$$iZ_2^{-1/2}\int d^4y \bar{V}_{\bar{p}\,\bar{s}}(y)(i\not\nabla_y - m)<\alpha\ out|W_{QED}^{-1}\psi(X(y))W_{QED}|\beta\ in> \qquad (7.75)$$

2. Reduction of a particle from an out-state

$$-iZ_2^{-1/2}\int d^4y \bar{U}_{p's'}(y)(i\not\nabla_y - m)<\alpha\ out|W_{QED}^{-1}\psi(X(y))W_{QED}|\beta\ in> \qquad (7.76)$$

3. Reduction of an anti-particle from an out-state

$$iZ_2^{-1/2}\int d^4y <\alpha\ out|W_{QED}^{-1}\bar{\psi}(X(y))W_{QED}|\beta\ in>(-i\overleftarrow{\not\nabla}_y - m)V_{\bar{p}'\,\bar{s}'}(y) \qquad (7.77)$$

where

$$V_{ps}(y) = \{m/[(2\pi)^3 E_p]\}^{1/2} v(p,s)e^{ip.y} \tag{7.78}$$

Electromagnetic Field

The LSZ reduction of a photon from an S matrix element:

$$S_{\alpha\beta k\lambda} = <\alpha \text{ out}|\beta\ \gamma(k\lambda) \text{ in}> \tag{7.79}$$

begins with

$$S_{\alpha\beta k\lambda} = <\alpha - \gamma(k\lambda) \text{ out}|\beta \text{ in}>$$
$$- iZ_3^{-1/2}\int d^3y\ A_{k\lambda}^{\mu *}(y)<\alpha \text{ out}|[A_{in}^{\mu}(y) - A_{out}^{\mu}(y)]|\beta \text{ in}> \tag{7.80}$$

where Z_3 is a normalization constant and

$$A_{k\lambda}^{\mu}(y) = [(2\pi)^3\omega_k]^{-1/2}e^{-ik.y}\ \varepsilon^{\mu}(k, \lambda) \tag{7.81}$$

Using

$$A^{\mu}(X(y)) \rightarrow \surd Z_3\ A_{in}^{\mu}(X(y)) \quad \text{as} \quad y^0 \rightarrow -\infty \tag{7.82}$$

$$A^{\mu}(X(y)) \rightarrow \surd Z_3\ A_{out}^{\mu}(X(y)) \quad \text{as} \quad y^0 \rightarrow \infty \tag{7.83}$$

and

$$A_a^{\mu}(y) = W_{QEDa}^{-1}(y)A_a^{\mu}(X(y))W_{QEDa}(y) \tag{7.84}$$

where a = "in" or "out" we arrive at the reduction expression following steps that parallel the scalar field reduction:

$$S_{\alpha\beta k\lambda} = <\alpha - \gamma(k\lambda) \text{ out}|\beta \text{ in}>$$
$$- iZ_3^{-1/2}\int d^4y\ A_{k\lambda}^{\mu *}(y)\Box_y<\alpha \text{ out}|W_{QED}^{-1}A^{\mu}(X(y))W_{QED}|\beta \text{ in}> \tag{7.85}$$

Time Ordered Products and Perturbation Theory

By repeated application of the LSZ procedure outlined above an S matrix element is reduced to the vacuum expectation value of time ordered products of fermion and photon fields.

$$<\alpha \text{ out}|\beta \text{ in}> = \ldots <0|T(\ldots W^{-1}(y_m)U^{-1}(y_m{}^0)\psi_{in}(X(y_m))U(y_m{}^0)W(y_m) \ldots W^{-1}(y_n)U^{-1}(y_n{}^0)\bar{\psi}_{in}(X(y_n))U(y_n{}^0)W(y_n) \ldots W^{-1}(y_o)U^{-1}(y_o{}^0)A_{in}{}^{\mu}(X(y_o))U(y_o{}^0)W(y_o) \ldots)|0> \ldots \tag{7.86}$$

Following the same development of the U matrix as described in chapter 6 with minor changes in details leads to the time ordered product for S matrix elements:

$$\tau(y_1,\ldots, y_n) = \frac{<0|T(\ldots\psi_{in}(X(y_m))\ldots\bar{\psi}_{in}(X(y_n))\ldots A_{in}{}^{\mu}(X(y_o))\ldots\exp[-i\int d^4y' \mathscr{H}_{FintQED}])|0>}{<0|T(\exp[-i\int d^4y' \mathscr{H}_{FintQED}])|0>} \tag{7.87}$$

where the QED interaction hamiltonian is

$$\mathscr{H}_{FintQED} = e_0{:}\bar{\psi}_{in}(X(y))\not{A}_{in}(X(y))\psi_{in}(X(y)){:} \tag{7.88}$$

plus mass counter terms.

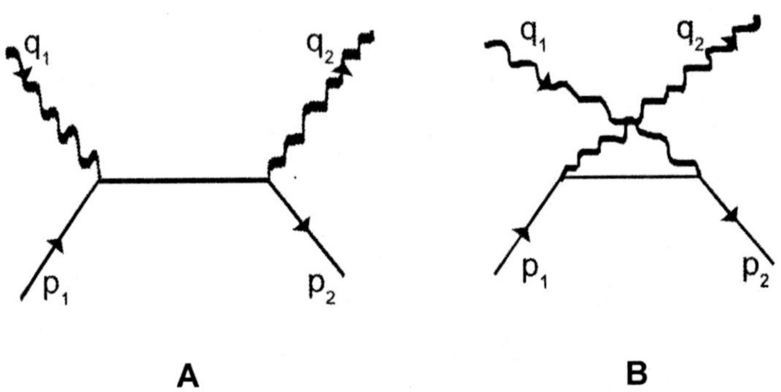

Figure 7.1. Lowest order elastic photon-electron scattering diagrams.

Example - Photon-Electron Elastic Scattering

In order to illustrate perturbative calculations in two-tier QED we will calculate the lowest order photon-electron elastic scattering S matrix element (Fig. 7.1).

We will see that at large distances (where the momenta are $\ll M_c$) the result is the same as the conventional QED calculation. At short distances (where the momenta are $\gg M_c$) the result differs markedly due to the effects of the Y field.

The S matrix element containing the contribution of this diagram is

$$S_{\gamma e} = (Z_2Z_3)^{-1}e_0^2 \prod_{j=1}^{6}\int d^4y_j\, \bar{U}_{p_2s_2}(y_4)A_{k_1\lambda_1}{}^{\mu_1}(y_2)A_{k_2\lambda_2}{}^{\mu_2*}(y_3)\Box_{y_2}\Box_{y_3}$$

$$(i\overrightarrow{\not\nabla}_{y_4} - m)\langle 0|T(\bar{\psi}_{in}(X(y_1))\, \psi_{in}(X(y_4))A_{in}{}^{\mu_1}(X(y_2))A_{in}{}^{\mu_2}(X(y_3))$$

$$:\bar{\psi}_{in}(X(y_5))\not{A}_{in}(X(y_5))\psi_{in}(X(y_5))::\bar{\psi}_{in}(X(y_6))\not{A}_{in}(X(y_6))\psi_{in}(X(y_6)):$$

$$)|0\rangle(-i\overleftarrow{\not\nabla}_{y_1} - m)U_{p_1s_1}(y_1) \qquad (7.89)$$

Diagram A in Fig. 7.1 corresponds to the Wick expansion:

$$S_{\gamma eA} = (Z_2Z_3)^{-1}e_0^2 \prod_{j=1}^{6}\int d^4y_j\, A_{k_1\lambda_1}{}^{\mu_1}(y_2)A_{k_2\lambda_2}{}^{\mu_2*}(y_3)\bar{U}_{p_2s_2}(y_4)\{$$

$$(i\overrightarrow{\not\nabla}_{y_4} - m)\, iS_F{}^{TT}(y_4 - y_6)\, \gamma^\nu\, iS_F{}^{TT}(y_6 - y_5)\gamma^\mu\;\; iS_F{}^{TT}(y_5 - y_1)(-i\overleftarrow{\not\nabla}_{y_1} - m)\}$$

$$U_{p_1s_1}(y_1)\, \Box_{y_2}\Box_{y_3}\, iD_F{}^{trTT}(y_5 - y_2)_{\mu_1\mu}iD_F{}^{trTT}(y_3 - y_6)_{\mu_2\nu}$$

$$(7.90)$$

After some manipulation eq. 7.90 can be placed in the form:

$$S_{\gamma eA} = (Z_2Z_3)^{-1}e_0^2\, (2\pi)^4\delta^4(p_2 + k_2 - p_1 - k_1)\bar{u}(p_2,s_2)\mathscr{S}_L{}^{TT}(p_2)\, \not{\epsilon}(k_2,\lambda_2)$$

$$\mathscr{S}^{TT}(p_1 + k_1)\, \not{\epsilon}(k_1,\lambda_1)\, \mathscr{S}_R{}^{TT}(p_1)\, u(p_1,s_1)\, \mathscr{D}^{TT}(k_1)\mathscr{D}^{TT}(k_2) \quad (7.91)$$

where

$$\mathscr{S}_L^{TT}(p) = iN_{fp} \int d^4y\, e^{ip.y}\, (i\not{\nabla}_y - m)\, S_F^{TT}(y) \qquad (7.92)$$

$$\mathscr{S}^{TT}(p) = i \int d^4y\, e^{ip.y}\, S_F^{TT}(y) \qquad (7.93)$$

$$\mathscr{S}_R^{TT}(p) = iN_{fp} \int d^4y\, e^{ip.y}\, S_F^{TT}(y)\, (i\overleftarrow{\not{\nabla}_y} - m) \qquad (7.94)$$

$$\mathscr{D}^{TT}(k) = iN_{\gamma k} \int d^4y\, e^{ik.y}\, \Box_y\, D_F^{trTT}(y)_{00} \qquad (7.95)$$

with

$$N_{fp} = \{m/[(2\pi)^3 E_p]\}^{1/2} \qquad (7.96)$$

and

$$N_{\gamma k} = [(2\pi)^3 \omega_k]^{-1/2} \qquad (7.97)$$

The factors $\mathscr{S}_L^{TT}(p_2)$, $\mathscr{S}_R^{TT}(p_1)$, and $\mathscr{D}^{TT}(k_1)$ and $\mathscr{D}^{TT}(k_2)$ serve to change the normalization of the in and out particle legs. They are a consequence of the dressing of the legs by Y particle "clouds." At long distance (low momentum) they approach the corresponding values of conventional QED:

$$\mathscr{S}_L^{TT}(p) \rightarrow iN_{fp} \qquad (7.98)$$

$$\mathscr{S}_R^{TT}(p) \rightarrow iN_{fp} \qquad (7.99)$$

$$\mathscr{D}^{TT}(k) \rightarrow iN_{\gamma k} \qquad (7.100)$$

if $p, k \ll M_c$.

External Leg "Normalizations"

The external leg factors $\mathcal{S}_L^{TT}(p)$, $\mathcal{S}_R^{TT}(p)$ and $\mathcal{D}^{TT}(k)$ change the normalization of the external legs due to the Y particle "cloud" surrounding each particle. In order find their form for large momenta we will first evaluate the large momentum limit of the fermion propagator as $z^2 \to 0$ (the light cone). Starting from eq. 7.45 one can show that (space-like limit)

$$iS_F^{TT}(z) \to \gamma^0 \in(z^0) M_c^3 [-M_c^2 \pi z^2/2]^{3/2} + \mathcal{O}(z^5) \qquad (7.101)$$

where $z^2 = z_0^2 - \mathbf{z}^2$. Therefore on dimensional grounds we see that

$$\mathcal{S}^{TT}(p) \backsim \gamma^0 M_c^6 p^{-7} + \mathcal{O}(p^{-9}) \qquad (7.102)$$

as p gets very large ($p \gg M_c$). At low momenta ($p \ll M_c$) the standard momentum space form of the fermion propagator is found:

$$\mathcal{S}^{TT}(p) \backsim (\not{p} - m + i\epsilon)^{-1} \qquad (7.103)$$

Similarly we find

$$\mathcal{S}_L^{TT}(p) \backsim M_c^6 p^{-6} + \mathcal{O}(p^{-8}) \qquad (7.104)$$

$$\mathcal{S}_R^{TT}(p) \backsim M_c^6 p^{-6} + \mathcal{O}(p^{-8}) \qquad (7.105)$$

as p gets very large ($p \gg M_c$), and

$$\mathcal{D}^{TT}(k) \backsim M_c^4 k^{-4} \qquad (7.106)$$

from eqs. 7.47 and 4.24 as k gets very large ($k \gg M_c$).

For small momenta compared to M_c we find the usual normalization:

$$\mathcal{S}_L^{TT}(p) = \mathcal{S}_R^{TT}(p) = iN_{fp} \qquad (7.107)$$

$$\mathcal{D}^{TT}(k) = iN_{\gamma k} \qquad (7.108)$$

Thus for low momenta (p, k $\ll M_c$) $S_{\gamma eA}$ yields the standard result of QED while at large momenta (p, k $\gg M_c$) we see a high power of inverse momentum showing the well behaved nature of the theory at short distances.

Renormalization of Two-Tier QED

Two-tier QED is a finite, unitary quantum field theory. The degree of divergence of a Feynman diagram term in *conventional QED* is

$$D = 4k - 2b - f \tag{7.109}$$

where

k = the number of internal momentum integrations
b = the number of internal photon lines
f = the number of internal electron lines

Many diagrams are thus divergent in QED and a renormalization program must be followed to achieve a theory with all divergences formally absorbed into renormalizations of the fundamental parameters of the theory. Despite the success of this approach and the excellent agreement of QED with experiment the presence of divergences in QED is logically unsatisfactory and suggests QED, and its successors Electroweak Theory and the Standard Model, are at best interim theories. Numerous attempts have been made to modify QED in order to eliminate its divergences. Some noteworthy attempts include the Lee-Wick formulation of QED and the Johnson-Baker-Willey model of QED. None of these attempts have succeeded for one reason or another.[31]

The degree of divergence formula has a different result in two-tier QED. It shows there are no divergences in two-tier QED. Thus it would be more aptly named the "degree of convergence." The formula is:

$$D^{TT} = 4k - 6b - 7f \tag{7.110}$$

with k, b and f as above. The coefficient of b is 6 in two-tier QED because the two-tier photon propagator behaves as k^{-6} at high momentum (see eq. 7.47 for the photon propagator and eq. 4.25 that shows its high momentum behavior). The coefficient of f is 7 in two-tier QED because the two-tier fermion propagator behaves as k^{-7} at high momentum (see eq. 7.102).

[31] Lee, T. D. and Wick, G. C., Phys. Rev. **D2**, 1033 (1970); T. D. Lee and G. C. Wick, Nucl. Phys. **B9**, 209 (1969) and references therein; M. Baker and K. Johnson, Phys. Rev. **D3**, 2516 (1971); M. Baker and K. Johnson, Phys. Rev. **D8**, 1110 (1973); K. Johnson, M. Baker, and R. Willey, Phys. Rev. **136**, B1111 (1964); K. Johnson, R. Willey, and M. Baker, Phys. Rev. **163**, 1699 (1967); S. Blaha,, Phys. Rev. **D9**, 2246 (1974) and references therein; S. Adler, Phys. Rev. **D5**, 3021 (1972).

We note the degree of Divergence in Two-Tier QED is always negative – indicating a finite theory!

$$D^{TT} < 0 \qquad (7.111)$$

For example the degree of divergence in two-tier QED of the lowest order in α (see Fig. 7.2): i) vacuum polarization diagram is $D^{TT} = -10$, ii) fermion self-energy is $D^{TT} = -9$ and iii) electromagnetic vertex correction is $D^{TT} = -16$.

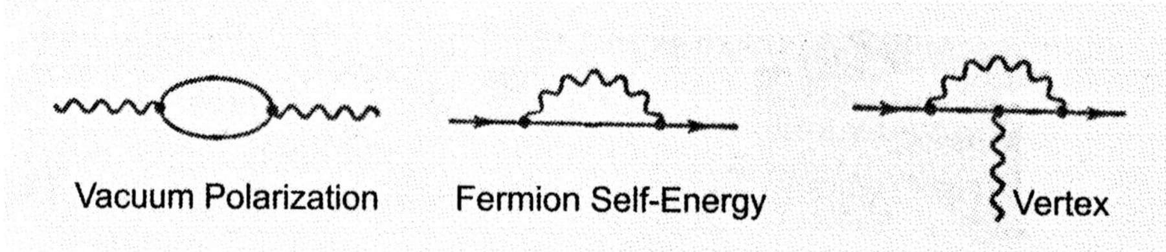

Figure 7.2 Some low order (normally divergent) diagrams for the vacuum polarization, fermion self-energy and electromagnetic vertex correction.

It is easy to see that all two-tier QED Feynman diagrams are ultra-violet finite.

Unitarity of Two-Tier QED

The remaining major issue is unitarity. Two-tier QED is unitary. The argument showing unitarity parallels the discussion of unitarity for scalar ϕ^4 quantum field theory in chapter 6.

Two-tier QED *superficially* also appears to have a unitarity problem due to the non-hermitean nature of its hamiltonian. The lack of hermiticity is entirely due to the appearance of iY^μ in the X^μ field coordinates.

Thus the interaction hamiltonian in eq. 7.88 is not hermitean:

$$H_{FintQED} = \int d^3y' \mathscr{H}_{FintQED}(\bar{\psi}_{in}(y' + iY(y')/M_c^2), A_{in}^\mu(y' + iY(y')/M_c^2), \psi_{in}(y' + iY(y')/M_c^2)) \qquad (7.112)$$

and

$$H_{FintQED}^\dagger = \int d^3y' \mathscr{H}_{FintQED}(\bar{\psi}_{in}(y' - iY(y')/M_c^2), A_{in}^\mu(y' - iY(y')/M_c^2), \psi_{in}(y' - iY(y')/M_c^2)) \qquad (7.113)$$

$$\neq H_{FintQED}$$

The metric operator (eq. 7.73) establishes the relation between $H_{FintQED}$ and its hermitean conjugate is

$$H_{FintQED} = V\, H_{FintQED}^{\dagger}\, V \tag{7.114}$$

Thus the Two-Tier QED S matrix is not unitary – S is pseudo-unitary in general:

$$S^{-1} = V\, S^{\dagger}\, V \tag{7.115}$$

Therefore

$$S^{\dagger} V S = V \tag{7.116}$$

We will now show that *the Two-Tier QED S matrix is effectively unitary between physical asymptotic states* which in two-tier QED are states consisting of charged fermions and photons only. The two-tier QED S matrix is *block diagonal* in the sense that physical in-states consisting only of charged fermions and photons will only produce physical out-states consisting of charged fermions and photons only. The consequence of block diagonality is that the *Two-Tier QED S matrix is unitary within the physical states sector.* To prove this point, consider eq 7.116 between physical states – each consisting of charged fermions and photons only.

$$<f\ in|V|i\ in> = <f\ in|S^{\dagger}\, VS|i\ in> = \sum_{n,\, m} <f\ in|S^{\dagger}\, |n><n|V|m><m|S|i\ in> \tag{7.117}$$

where the block diagonality of S, and the diagonality of V, limits the intermediate states |n> and |m> to states containing charged fermions and photons only. We note V is diagonal with eigenvalue one between states consisting of charged fermions and photons only. Thus eq. 7.117 becomes

$$\delta_{fi} = \sum_{n} <f\ in|S^{\dagger}\, |n><n|S|i\ in> \tag{7.118}$$

or

$$S_{physQED}^{\dagger}\, S_{physQED} = I$$

which is equivalent to

$$S_{physQED}^{\dagger} = S_{physQED}^{-1} \tag{7.119}$$

proving unitarity in the physical states sector – the states consisting of charged fermions and photons only.

Comparison with String and Superstring Theories

A pseudo-unitary S matrix (with some states of positive metric and others of negative metric) is a common feature of string and superstring theories. These theories, like Two-Tier QED, define a set of physical, positive metric states whose S matrix elements satisfy the unitarity requirement. Thus their S matrix has a form of block diagonality which enables them to have a "physical" S matrix in a manner analogous to Two-Tier QED.

Correspondence Principle for Two-Tier QED

The two-tier QED theory that we have developed has the behavior of conventional QED at "low energies." Since M_c sets the scale at which deviations from the normal QED results become significant we can set M_c to an extremely high value such as the Planck mass and obtain as close an agreement with conventional QED as we wish. Thus two-tier QED implements a type of Correspondence Principle with conventional QED as its "low energy" limit.

Consistency with Precision QED Experiments

Since we can obtain results as close to conventional QED with a sufficiently high choice of M_c, two-tier QED has the same excellent agreement[32] with experiment as conventional QED.

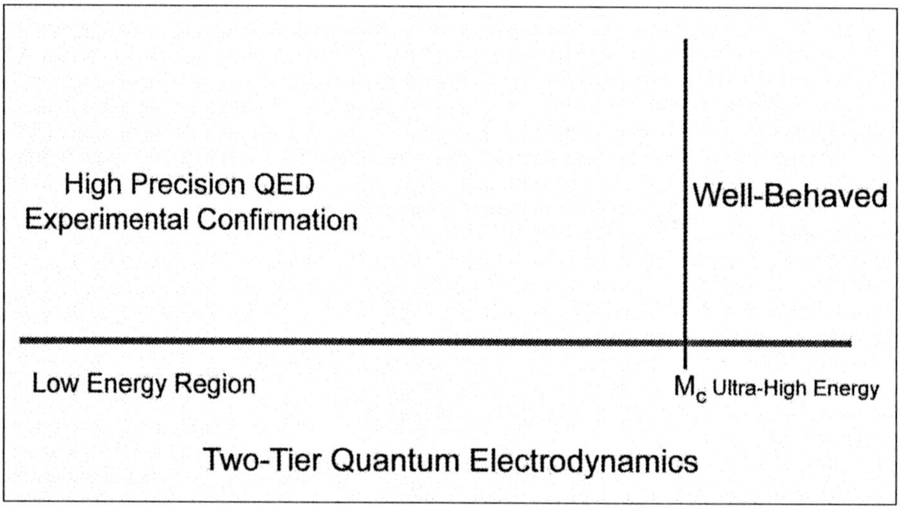

Figure 7.3. A Schematic View of Two-Tier QED.

[32] See, for example, T. Kinoshita, "The Fine Structure Constant", Cornell University preprint CLNS 96/1406 (1996); V. W. Hughes and T. Kinoshita, Rev. Mod. Phys. **71**, S133 (1999) and references therein.

Calculational Rules for Feynman Diagrams

This section describes the procedure for the calculation of S matrix terms corresponding Feynman diagrams in Two-Tier QED and other quantum field theories. By design the approach parallels the perturbative approaches of conventional QED and other quantum field theories. The examples considered earlier in this chapter, and the preceding chapter, illustrate two-tier perturbation theory.

Procedure: Follow the conventional procedure to form the expression for each Feynman diagram in the perturbative calculation. Then replace each propagator in each expression with the corresponding two-tier propagator as specified below. Then evaluate the expression.

Feynman Propagators and their Two-Tier Propagator Equivalents

Two-tier propagators are denoted with the superscript "TT".

<u>Spin 0 Propagator Case</u>

$$\Delta_F(p) = (p^2 - m^2 + i\varepsilon)^{-1} \qquad \rightarrow \int d^4z\, e^{+ip\bullet z}\, \Delta_F^{TT}(z) \qquad (7.120)$$

where $\Delta_F^{TT}(z)$ is given by eq. 4.9.

<u>Spin 1 Photon Propagator Case</u>

$$D_F(p)_{\mu\nu} = -g_{\mu\nu}(p^2 + i\varepsilon)^{-1} \qquad \rightarrow \int d^4z\, e^{+ip\bullet z}\, D_F^{TT}(z)_{\mu\nu} \qquad (7.121)$$

where $D_F^{TT}(z)_{\mu\nu}$ is given by eq. 7.47.

<u>Spin ½ Fermion Propagator Case</u>

$$S_F(p) = (\not{p} - m + i\varepsilon)^{-1} \qquad \rightarrow \qquad \int d^4z\, e^{+ip\bullet z}\, S_F^{TT}(z) \qquad (7.122)$$

where $S_F^{TT}(z)$ is given by eq. 6.88.

<u>Spin 2 Massless Boson (graviton) Propagator Case</u>

$$\Delta_{F2}(p)_{\mu\nu\rho\sigma} = b_{\mu\nu\rho\sigma}(p^2 - m^2 + i\varepsilon)^{-1} \rightarrow \int d^4z\, e^{+ip\bullet z}\, \Delta_{F2}^{TT}(z)_{\mu\nu\rho\sigma} \qquad (7.123)$$

where $\Delta_{F2}^{TT}(z)_{\mu\nu\rho\sigma}$ the graviton propagator.

Two-Tier Coulomb Potential vs. Conventional Coulomb Potential

The familiar Coulomb potential is (for two particles of opposite unit electric charge):

$$V_{\mathrm{Coulomb}} = -\,a/|\mathbf{r}| \tag{7.124}$$

The Two-Tier QED Coulomb potential (eq. 7.49) is:

$$V_{\mathrm{Two\text{-}TierCoul}} = -\,a\Phi(M_c^{\,2}\pi|\mathbf{r}|^2)/|\mathbf{r}| \tag{7.125}$$

where $\Phi(x)$ is the error function.[33] At small distances ($\pi r^2 \ll M_c^{-2}$)

$$V_{\mathrm{Two\text{-}TierCoul}} \rightarrow -\,a2\sqrt{\pi}\,M_c^{\,2}\,|\mathbf{r}| \tag{7.126}$$

a linear potential, and at large distances ($\pi r^2 \gg M_c^{-2}$)

$$V_{\mathrm{Two\text{-}TierCoul}} \rightarrow V_{\mathrm{Coulomb}} \tag{7.127}$$

The Two-Tier Coulomb potential has a minimum at

$$M_c^{\,2}\pi|\mathbf{r}|^2 = 1 \tag{7.128}$$

At the minimum $V_{\mathrm{Two\text{-}TierCoul}}$ has the value:

$$V_{\mathrm{Two\text{-}TierCoulMIN}} = -\,.8427a\sqrt{\pi}\,M_c \tag{7.129}$$

If we define distance in terms of the mass scale

$$|\mathbf{r}| = R/M_c \tag{7.130}$$

then

$$V_{\mathrm{Two\text{-}TierCoul}} = -\,aM_c\Phi(\pi R^2)/R \tag{7.131}$$

Fig. 7.4 displays a plot of $V_{\mathrm{Two\text{-}TierCoul}}/aM_c$ showing the general form of the Two-Tier Coulomb potential.

---

[33] W. Magnus and F. Oberhettinger, *Formulas and Theorems for the Special Functions of Mathematical Physics* (Chelsea Publishing Co., New York, 1949) page 96.

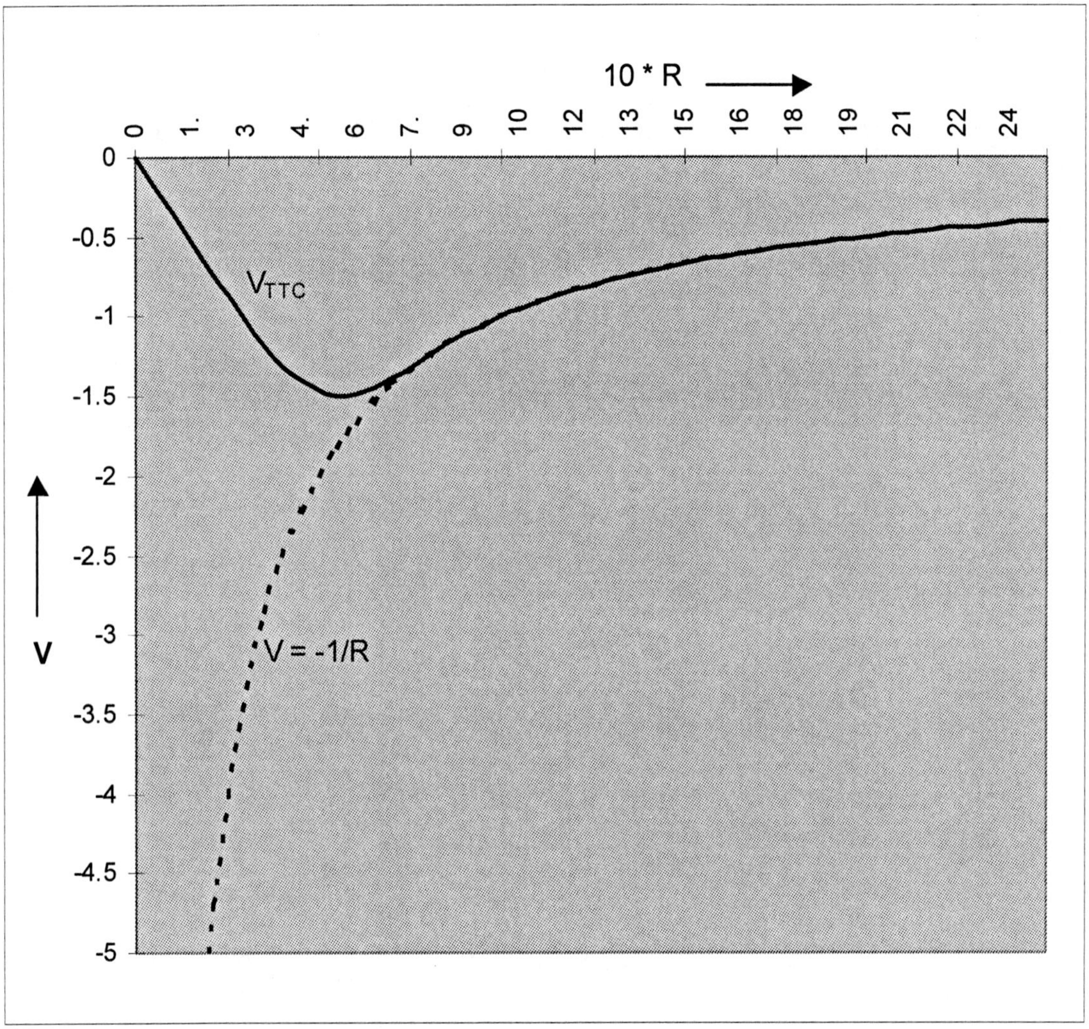

Figure 7.4. Plot of Two-Tier Coulomb potential divided by αM_c: $V_{TTC} = V_{Two\text{-}TierCoul}/\alpha M_c$. V_{TTC} is dimensionless while R is the radial distance measured in units of M_c^{-1}.

Doubly-Charged Dilepton Resonances

Doubly charged dilepton resonances such as e^-e^- or $\mu^-\mu^-$ resonances are possible since it appears that the repulsive electromagnetic force between similarly charged leptons goes to zero as the distance between them goes to zero creating a potential barrier that might, in principle, lead to binding.

However because the distance between the particles at which the repulsive Coulomb potential starts to decline (Fig. 7.5) is very small, a broad resonance is the best that one can expect. Tunneling will cause rapid decay of the bound state.

Note that the peak of the potential barrier, for $M_c = 10^5$ GeV/c^2, is at $r = 5.6 \times 10^{-6}$ GeV^{-1} $\equiv 1.1 \times 10^{-21}$ m (by eq. 7.128) while the electron Compton wavelength is 3.861×10^{-13} m. Thus the electron's Compton wavelength is roughly 4×10^8 times larger than the distance of the maximum of the Two-Tier Coulomb potential. As a result the electrons in a dilepton resonance would immediately tunnel out of the binding region (*zitterbewegung*!) despite the strength of the potential at maximum: 1.09×10^5 GeV according to eq. 7.129. Even if one could create a resonance state of two τ^- leptons with τ Compton wavelengths of 1.135×10^{-16} m the τ wavelength would be 10^5 times larger than the distance of the minimum of the two-tier Coulomb potential if $M_c = 10^5$ GeV.

If M_c is larger than 10^5 GeV then the location of the peak of the potential is even smaller. Thus a dilepton resonance can be expected to be very broad and very short-lived.

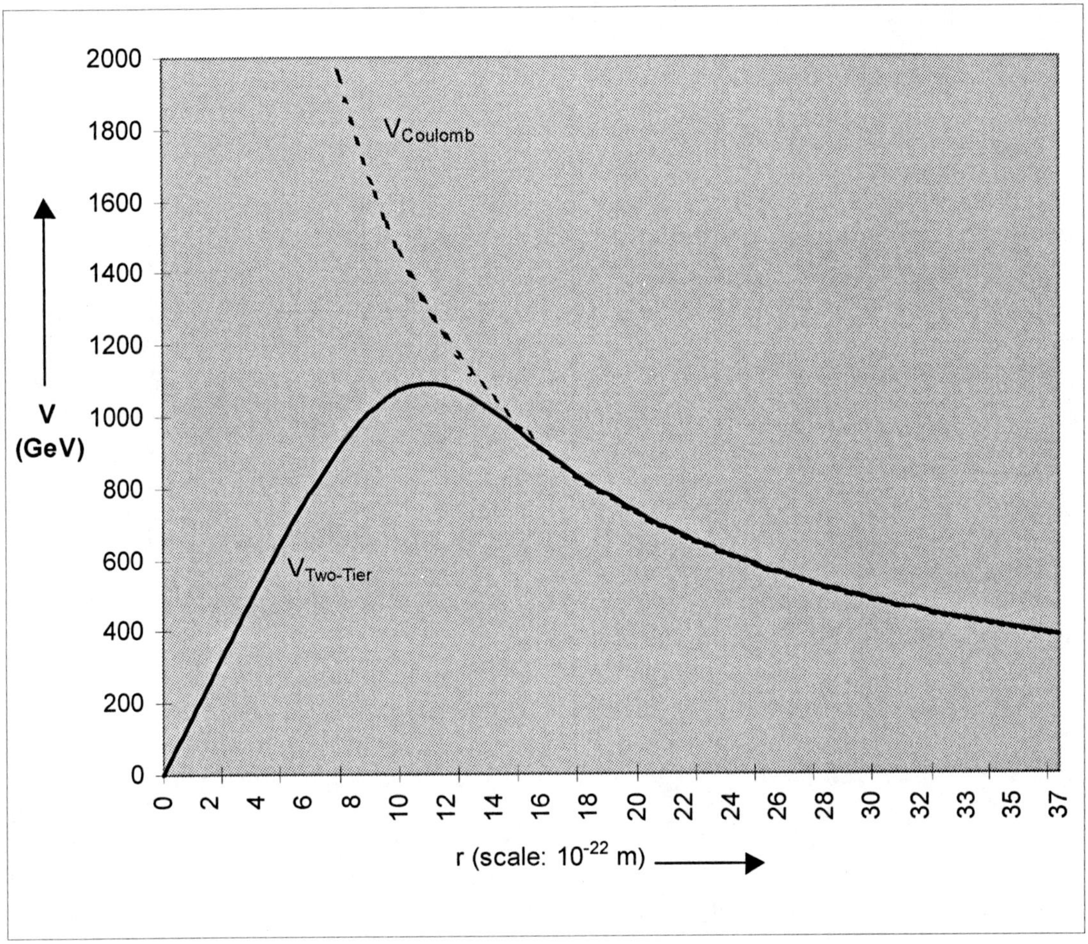

Figure 7.5. Two-Tier Coulomb Potential compared to conventional Coulomb potential for $M_c = 10^5$ GeV.

8. Two-Tier Standard Model Theory

Introduction

The Standard Model is a unified model of the electromagnetic, weak and strong interactions. It is formed from Electroweak Theory and an SU(3) color gauge theory of the strong interaction called Quantum Chromodynamics (QCD). Each part is separately renormalizable, as is the combined theory. This ad hoc theory of all known interactions except gravitation has been remarkably successful in accounting for the experimental data at energies currently accessible at accelerators. There are a number of phenomena that are not understood within the framework of the Standard Model such as dark matter and dark energy, some CP violation phenomena, and the spin dependence observed in certain experiments. There is also an appreciation of the ad hoc nature of the unification of the strong and electroweak sectors. They are "just glued together". An overall symmetry encompassing both sectors (broken as it would be) would be more reassuring and more elegant. However the theory works, and appears to work well, for most particle physics phenomena.

Superstring theory has become a seriously considered alternative to the Standard Model. It appears to offer a finite theoretical framework without divergences. It offers the hope of a rational grand unification. It is mathematically interesting and elegant. However, it has almost no experimental support. It predicts many particles that remain to be found. It requires additional dimensions that are not evidenced. It does not have a demonstrable, satisfactory "low energy limit" that resembles the Standard Model. It has the problem of too many possibilities: after thirty years of effort one cannot point to a Superstring theory and say it is THE Superstring theory. Considering that far more theoretical effort has been expended on Superstring theory by more theorists then on any other theory of physics, its physical justification, its nature, and its contact with experiment is still not well understood.

In the hope of developing a sound theory that clearly makes contact with known experimental data, that has a direct connection to the Standard Model, that is finite (no divergences), and that can be directly extended to encompass gravitation, we have developed a two-tier formulation of the Standard Model and its variants. It is admittedly far more modest in its scope than Superstring theory. Yet it allows us to have a finite theory – a major goal of Superstring theory – and to address cosmological issues by its easy unification with two-tier quantum gravity (next chapter) – also a finite theory without divergences.

Thus we have a theory that accounts for all known interactions and yet incrementally extends known successful theories – in a significant way. With (perhaps) the exception of

General Relativity, the growth of particle physics theory in this century has been largely incremental.

The nature of the mechanism that two-tier quantum field theory uses to avoid divergences has the important feature that it does not depend on "magic cancellations", does not depend on details of the symmetries, and does not depend on details of the form of the theories. Thus all the variations and extensions of the Standard Model have an equivalent two-tier version that is finite. This feature gives the theorist the flexibility to modify and extend the two-tier Standard Model without worrying about the introduction of new divergences.

The other major advantage is that the theory is very much like the world we see at current energies – four dimensions and no large number of new particles. It leaves open the door for new interactions and new particles IF they are found at higher energies. Thus it is not a showstopper. Nor is it a prediction of a vast desert extending to the highest energies. It gives a finite, unified theory with an "Open Door" policy for the future.

Formulation of Two-Tier Standard Models

There are many excellent texts[34] on the Standard Model. Therefore this author sees no reason to recapitulate the features of the Standard Model. We will focus on the form and features of a generic two-tier version of the Standard Model.

We will assume that the form of the Standard Model Lagrangian is:

$$\mathscr{L}_{SM} = J\mathscr{L}_F{}^{SM}(X^\mu) + \mathscr{L}_C(X^\mu(y), \partial X^\mu(y)/\partial y^\nu) \qquad (8.1)$$

where $\mathscr{L}_F{}^{SM}$ is the complete "normal" quantum field theory Lagrangian for the Standard Model variant under consideration and J is a Jacobian (eq. 2.21). *All particle fields in* $\mathscr{L}_F{}^{SM}$ *are assumed to be functions of the* X^μ *coordinate only. The dependence of the particle fields on the "underlying" coordinates* y^μ *is assumed to be solely through* X^μ. The Lagrangian $\mathscr{L}_{SM}$ is a separable Lagrangian of the type of eq. 2.26 embodying the composition of extrema described in chapter 2. Scalar particle examples of separable Lagrangians appear in eqs. 2.96, 3.10 and 3.15. A scalar particle ϕ^4 theory with a separable Lagrangian is specified by eqs. 5.20 – 5.24 and decribed in detail in chapters 5 and 6. The separable Lagrangian two-tier version of QED is specified by eqs. 7.1 – 7.5 and described in some detail in chapter 7.

In all of these cases we have specified the coordinate part of the Lagrangian $\mathscr{L}_C$ as

[34] Some of the many excellent texts: P. Ramond, *Journeys Beyond the Standard Model* (Westview Press, New York, 1999); W. N. Cottingham and D. A. Greenwood, *An Introduction to the Standard Model of Particle Physics* (Cambridge University Press, New York, 1999); K. Huang, *Quarks, Leptons and Gauge Fields* (World Scientific, River Edge, NJ 1992); T. P, Cheng and L. F. Li, *Gauge Theories of the Elementary Particles* (Oxford University Press, New York, 1988); J. Donoghue, E. Golowich and B. Holstein, *Dynamics of the Standard Model* (Cambridge University Press, New York, 1994).

$$\mathscr{L}_C = +\tfrac{1}{4} M_c^4 F^{\mu\nu} F_{\mu\nu} \tag{3.10}$$

with

$$F_{\mu\nu} = \partial X_\mu/\partial y^\nu - \partial X_\nu/\partial y^\mu \tag{3.11}$$

$$X_\mu(y) = y_\mu + i\, Y_\mu(y)/M_c^2 \tag{3.12}$$

We then re-expressed the Lagrangian as

$$\mathscr{L}_C = -\tfrac{1}{4} F_Y^{\mu\nu} F_{Y\mu\nu} \tag{3.15}$$

with

$$F_{Y\mu\nu} = \partial Y_\mu/\partial y^\nu - \partial Y_\nu/\partial y^\mu \tag{3.14}$$

and developed the corresponding quantum field theory using Y^μ. The field equations of the theory are found using the conventional approach. The Hamiltonian, energy-momentum tensor and conserved quantities are also found using a conventional approach as illustrated in chapters 2, 3 and 4. The canonical quantization procedure is also followed. Thus the application of these procedures to the Standard Model will yield the standard results with the sole difference being that all fields are functions of X^μ and all derivatives of fields are derivatives with respect to X^μ. A perturbation theory along the lines of chapters 5 and 6 can then be directly developed.

The net result is that the free field Feynman propagators in a Feynman diagram approach to perturbative calculations are each replaced with the corresponding two-tier propagator. All coordinate space integrations are over the y coordinates.

Rule 1: *If the conventional coordinate space Feynman propagator for a particle has the form:*

$$G_{\ldots}(z) = \int d^4p\, e^{-ip\cdot z}\, G_{\ldots}(p)/(2\pi)^4 \tag{8.2}$$

where ... represents any space-time indices, spin indices and internal symmetry indices, then the equivalent coordinate space (y – coordinates) two-tier propagator is

$$G_{\ldots}^{TT}(z) = \int d^4p\, e^{-ip\cdot z}\, G_{\ldots}(p) R(p, z)/(2\pi)^4 \tag{8.3}$$

where R(p, z) is specified in eqs. 4.12 – 4.20.

The vertex expressions remain identically the same as in the conventional Standard Model. All integrations are done in the y-coordinate space as shown in the perturbation discussion leading to eq. 6.37 and the perturbation theory calculations at the end of chapter 6 and in chapter 7.

Procedure to Construct the Two-Tier Equivalent of a Conventional Feynman Diagram Expression

After we form the expression for a Feynman diagram in a conventional Standard Model variant, we can construct the Two-Tier equivalent by taking *every* particle or ghost ghost propagator factor of the form:

$$G\ldots(p)$$

in momentum space, and replacing it with

$$G\ldots^{TT}(p) = \int d^4z\, e^{ip\bullet z}\, G\ldots^{TT}(z)$$

$$= \int d^4z\, e^{ip\bullet z} \int d^4p'\, e^{-ip'\bullet z}\, G\ldots(p')R(p', z)/(2\pi)^4$$

The resulting expression is the two-tier expression for the Feynman diagram.

"Low Energy" Behavior of the Two-Tier Standard Model

Since $R(p, z) \cong 1$ for $p \ll M_c$ the two-tier Standard Model is virtually identical to the conventional, corresponding Standard Model variant for energies much less than M_c. Since M_c can be very large – perhaps equal to the Planck mass or larger, the two-tier Standard Model predictions at current energies can be made arbitrarily close to the "low energy" results of the corresponding Standard Model. The Standard Model is a limiting case of the two-tier Standard Model – thus implementing a *Correspondence Principle*.

"High Energy" Behavior of the Two-Tier Standard Model

At high energies (short distances) where p is of the order of M_c or larger, R(p, q) provides a Gaussian damping factor that makes all perturbation theory calculations ultra-violet finite. This has been shown in exhaustive detail for fermions and bosons in earlier chapters.

Massive Vector Bosons – No Divergence Problems

The only new case is the case of massive vector bosons. The propagator has the general form:

$$D_{VM}(k)_{\mu\nu} = -(g_{\mu\nu} - k_\mu k_\nu/m^2)(k^2 - m^2 + i\varepsilon)^{-1} \qquad (8.4)$$

in momentum space in conventional quantum field theory modulo internal symmetry indices. The two-tier quantum field theory equivalent is:

$$D_{VM}{}^{TT}(z)_{\mu\nu} = \int d^4k\, e^{-ip\bullet z}\, D_{VM}(k)_{\mu\nu} R(k, z)/(2\pi)^4 \tag{8.5}$$

in y-coordinate space. The leading momentum space behavior at high-energy of $D_{VM}{}^{TT}$ is generated by the $g_{\mu\nu}$ term (unlike the situation in conventional quantum field theory):

$$D_{VM}{}^{TT}(k)_{\mu\nu} \sim g_{\mu\nu} k^{-6} \tag{8.6}$$

as in eq. 6.101 for the massless vector boson case.

Parenthetically we note that the two-tier version of the ghost propagators appearing in the gauge theory sectors is also handled by Rule 1. These propagators have the same high energy leading momentum behavior as massless scalar bosons, k^{-6}.

Thus the leading high-energy behavior of all two-tier propagators of Standard Model particles and ghosts is a high negative power of momentum. As a result all perturbative calculations are ultra-violet finite.

Path Integral Formulation

Until now we have been viewing quantum field theory calculations in terms of conventional, Feynman diagram-based, perturbation theory. The appearance of a number of internal symmetries in the Standard Model (usually SU(2)⊗SU(1)⊗SU(3)) that are implemented with Yang-Mills gauge fields often makes it convenient to use a path integral formalism. Since there are many excellent introductions to the path integral formalism in relation to Yang-Mills fields and the Standard Model we will confine our discussion to aspects specifically related to the two-tier formalism for quantum field theory that we have been developing.

We will consider the case of a two-tier scalar particle quantum field theory initially for the sake of simplicity, and then consider two-tier Yang-Mills gauge field theories. The primary quantity in the path integral formulation is

$$Z(J) = \langle 0^+|0^-\rangle \tag{8.8}$$

$$= N \int D^4Y D\phi \exp\{ i\int d^4y\, [\mathscr{L} + j_\mu(y)Y^\mu(y) + J(y)\phi(X)]\}|_{j_\mu = 0} \tag{8.9}$$

$$= \langle 0|T\big(\exp\{i\int d^4y\, [\mathscr{L} + j_\mu(y)Y^\mu(y) + J(y)\phi(X)]\}\big)|0\rangle \tag{8.10}$$

with X defined by eq. 3.12. Eq. 8.9 expresses the path integral as a product of integrals over classical field points, and eq. 8.10 provides an operator formulation of the path integral. N is a normalization factor. (As earlier we use units where $\hbar = 1$.) We note that $<0^+|0^->$ is the probability amplitude that the vacuum state at $y^0 = -\infty$ transitions to the vacuum state at $y^0 = +\infty$. The external sources $j_\mu(y)$ and $J(y)$ are arbitrary c-number functions of the respective variables with the restriction: as $y^0 \equiv Y^0 \rightarrow \pm\infty$ (in the Y Coulomb gauge that we are using) both external sources approach zero:

$$\lim_{y^0 \to \pm\infty} j_\mu(y) = \lim_{y^0 \to \pm\infty} J(y) = 0 \tag{8.11}$$

Notice $j_\mu(y)$ is set to zero after the evaluation of the path integral. By assumption there are no external sources, or in/out asymptotic states containing Y particles.

The functional derivatives satisfy

$$\delta J(y_1)/\delta J(y_2) = \delta^4(y_1 - y_2) \tag{8.12}$$

and

$$\delta j_\mu(y_1)/\delta j^\nu(y_2) = g_{\mu\nu}\delta^4(y_1 - y_2) \tag{8.13}$$

General Procedure

In the general case we will assume that the Lagrangian (with external sources added) has the form:

$$\mathscr{L} = \mathcal{J}(Y, y)[\mathscr{L}_{Fint} + \sum_i \mathscr{L}_{F0i}] + \mathscr{L}_C \tag{8.14}$$

where $\mathscr{L}_{Fint}$ is the "interaction Lagrangian" that contains all non-quadratic field operator product terms, $\mathscr{L}_{F0i}$ is the "free field" Lagrangian for particle species i, $\mathscr{L}_C$ is the (free) Lagrangian for the Y^μ abelian gauge field described in detail in earlier chapters, and $\mathcal{J}(Y, y)$ is the Jacobian for the transformation of y^μ coordinates to X^μ coordinates. (We now use script J for the Jacobian because the sources use the symbol J.) $\mathscr{L}_{Fint}$ and $\mathscr{L}_{F0i}$ are functions of the fields corresponding to the physical particles of the theory. Each field is solely a function of X^μ and all derivatives are with respect to X^μ as developed in the two-tier models of preceding chapters:

$$X_\mu(y) = y_\mu + i\,Y_\mu(y)/M_c^2 \tag{3.12}$$

In the present discussion we will deal directly with the Y^μ field and use eq. 3.12 as a notational convenience. We begin with eqs. 8.9 and 8.14 for the case of n scalar Klein-Gordon particles, which we write as:

$$Z(J) = N\,\{\exp\{\,i\int d^4y\,\mathcal{J}(-i\delta/\delta j^\nu(y)\,,\,y)\mathscr{L}_{\mathrm{Fint}}(-i\,\delta/\delta J_1(y),\,\ldots,\,-i\,\delta/\delta J_n(y))\}\cdot$$

$$\prod_k [\int D\phi_k \exp\{\,i\int d^4y\,[\mathcal{J}(-i\delta/\delta j^\nu(y),\,y)\mathscr{L}_{\mathrm{F0k}} + J_k(y)\phi_k(X)]\}|_{j_\mu = 0}]\cdot$$

$$\int D^4Y\,\exp\{\,i\int d^4y[\mathscr{L}_C \;+ j_\mu(y)Y^\mu(y)]\}\}|_{j_\mu = 0} \qquad (8.15)$$

using functional derivatives with respect to the sources. We note that we set $j^\mu = 0$ *after* evaluating each of the n free field factors in eq. 8.15. The physical justification for this procedure is that we wish the free field propagators to be truly free field propagators and not depend on external sources, and not be convoluted together via the Y^μ field. The two-tier nature of the quantum field theory requires this procedure. At the end of the evaluation of eq. 8.15 we set $j^\mu = 0$ since we require Y particles do not appear in in-states or out-states.

We will now evaluate eq. 8.15 in stages since a number of novel features appear in its evaluation although the path integral evaluations themselves are conventional. We start by performing the Y integration. The Y field is a free abelian gauge theory. We choose the same Coulomb gauge as in previous chapters $\nabla\cdot \mathbf{Y} = 0$ and $Y^0 = 0$. Consider

$$Z_Y(j_\mu) = \int D^4Y\,\exp\{\,i\int d^4y[\mathscr{L}_C + j_\mu(y)Y^\mu(y)]\} \qquad (8.16)$$

with $\mathscr{L}_C$ given by eq. 3.15. After some manipulations we find

$$Z_Y(j_\mu) = \exp\{\,-i\int d^4y_1 d^4y_2\; j^i(y_1)D_{Fij}(y_1 - y_2)j^j(y_2)/2\} \qquad (8.17)$$

with

$$D_{Fij}(y_1 - y_2) = \int d^4k\; e^{-ik\bullet(y_1 - y_2)}\,(\delta_{ij} - k_ik_j/\mathbf{k}^2)/[(2\pi)^4(k^2 + i\varepsilon)] \qquad (8.18)$$

the spatial components of the massless vector boson Feynman propagator in the Coulomb gauge.

We now evaluate one of the factors in the n free boson exponentiated Lagrangian:

$$Z_{\phi_k}(J_k) = \int D\phi_k \exp\{ i\int d^4y\, [\mathcal{J}(-i\delta/\delta j^{\nu}(y), y)\mathcal{L}_{F0k} + J_k(y)\phi_k(X)]\}|_{j_\mu=0} Z_Y(j_\mu) \tag{8.19}$$

The free Klein-Gordon Lagrangian is:

$$\mathcal{L}_{F0k} = ½\, [\, (\partial\phi_k/\partial X^{\nu})^2 - m_k^{\,2}\phi_k^{\,2}] \tag{8.20}$$

As a result the classical action can be written as

$$S_{0k}[\phi_k] = \int d^4y\, \mathcal{J}(-i\delta/\delta j^{\nu}(y), y)\mathcal{L}_{F0k} \tag{8.21}$$

$$= -½\int d^4X_{op}\; \phi_k(X_{op})(\Box_X + m_k^{\,2})\phi_k(X_{op}) \tag{8.22}$$

under the usual assumptions of good behavior at infinity and utilizing the Jacobian factor with

$$X_{op\mu}(y) = y_\mu + \; M_c^{-2}\, \delta/\delta j^{\mu}(y) \tag{8.23}$$

and with

$$\Box_X = (\partial/\partial X_{op}{}^{\nu})(\partial/\partial X_{op\nu}) \tag{8.24}$$

Inserting eq. 8.21 In eq. 8.19 we obtain:

$$Z_Y(j_\mu)Z_{\phi_k}(J_k) = \int D\phi_k \exp\{ i \int d^4X_{op}[-½\phi_k(X_{op})(\Box_X + m_k^{\,2})\phi_k(X_{op}) +$$

$$+ J_k(y)\phi_k(y)/\mathcal{J}(-i\delta/\delta j^{\nu}(y), y)]]\}|_{j_\mu=0} Z_Y(j_\mu) \tag{8.25}$$

The (Gaussian in ϕ_k) path integral in eq. 8.25 can be performed using standard techniques to yield:

$$Z_Y(j_\mu)Z_{\phi_k}(J_k) = \exp\{ -i \int d^4X_{op}(y_1)d^4X_{op}(y_2)[J_k(y_1)/\mathcal{J}(-i\delta/\delta j^{\nu}(y_1), y_1)]\cdot$$

$$\Delta_{Fk}{}^{TT}(y_1 - y_2)[J_k(y_2)/\mathcal{J}(-i\delta/\delta j^{\nu}(y_2), y_2)]/2\}|_{j_\mu=0} Z_Y(j_\mu)$$

$$= Z_Y(j_\mu)\exp\{-i\int d^4y_1 d^4y_2\, J_k(y_1)\Delta_{Fk}{}^{TT}(y_1 - y_2)J_k(y_2)/2\} \tag{8.26}$$

after changing the integration variables in the exponential using the Jacobian with $\Delta_{Fk}{}^{TT}(y_1 - y_2)$ given by eqs. 6.51-52. The added index k serves to identify the propagator $\Delta_{Fk}{}^{TT}(y_1 - y_2)$ as having the mass m_k.

The derivation of eq. 8.25 uses functional techniques in a straightforward way. We can derive the form of $\Delta_{Fk}{}^{TT}(y_1 - y_2)$ best in momentum space. The operator $(\Box_X + m_k^2)^{-1}$ can be represented via

$$(\Box_X + m_k^2)^{-1}g(X_{op}(y_1))Z_Y(j_\mu) = (\Box_X + m_k^2)^{-1}\int d^4p\{e^{-ip\bullet X_{op}(y_1)}g(p)/(2\pi)^4\}Z_Y(j_\mu)$$

$$= -\int d^4p\, \{e^{-ip\bullet X_{op}(y_1)}g(p)/[(2\pi)^4\, (p^2 - m_k^2 + i\varepsilon)]\}\, Z_Y(j_\mu)$$

$$= -\int d^4X_{op}(y_2)\Delta_{Fkop}{}^{TT}(X_{op}(y_1) - X_{op}(y_2))g(X_{op}(y_2))Z_Y(j_\mu)$$

$$= -\int d^4X_{op}(y_2)\Delta_{Fkop}{}^{TT}(X_{op}(y_1) - X_{op}(y_2))Z_Y(j_\mu)g(X(y_2))$$

We now calculate

$$\Delta_{Fkop}{}^{TT}(X_{op}(y_1) - X_{op}(y_2))Z_Y(j_\mu) = \int d^4p\, \{e^{-ip\bullet(y_1 - y_2)}/[(2\pi)^4\, (p^2 - m_k^2 + i\varepsilon)]\}\cdot$$

$$\exp(-ip^\nu[\delta/\delta j^\nu(y_1) - \delta/\delta j^\nu(y_2)]/M_c^2)\, Z_Y(j_\mu)$$

$$= Z_Y(j_\mu)\int d^4p\, \{e^{-ip\bullet(y_1 - y_2)}/[(2\pi)^4\, (p^2 - m_k^2 + i\varepsilon)]\}\cdot$$

$$\exp(-ip^i p^j[D_{Fij}(y_1 - y_2) + D_{Fij}(y_2 - y_1)]/(2M_c^4)) \tag{8.27}$$

where $D_{ij}(y_1 - y_2)$ is given by eq. 8.18. The last factor is the Gaussian damping factor that appears in two-tier propagators as repeatedly seen earlier – for example in eq. 6.53 and 6.54:

$$R(\mathbf{p}, y_1 - y_2) = \exp(-ip^i p^j[D_{Fij}(y_1 - y_2) + D_{Fij}(y_2 - y_1)]/(2M_c^4)) \tag{8.28}$$

Eq. 8.26 then follows with the two-tier scalar Klein-Gordon Feynman propagator.

Therefore the n scalar Klein-Gordon particle path integral of eq. 8.15 simplifies to

$$Z(J) = N \{\exp\{ i\int d^4y\, \mathcal{J}(-i\delta/\delta j^\nu(y), y)\, \mathscr{L}_{Fint}(-i\delta/\delta J_1(y), \ldots, -i\delta/\delta J_n(y))\}\cdot$$

$$\exp\{-\sum_k [\, i\int d^4y_1 d^4y_2\, J_k(y_1)\Delta_{Fk}{}^{TT}(y_1 - y_2)J_k(y_2)/2]\}\, Z_Y(j_\mu)\,\}|_{j_\mu = 0} \qquad (8.29)$$

The only dependence on functional derivatives with respect to j^μ in eq. 8.29 is in the Jacobian of the interaction Lagrangian term. It is readily seen that the Jacobian effectively reduces to one if there is no external physical source for Y particles (as we have assumed.) Consider

$$\mathcal{J}(-i\delta/\delta j^\nu(y),y)Z_Y(j_\mu)\,\Big|_{j_\mu = 0} = \{\varepsilon^{ijk}(\delta_{1i} + M_C{}^{-2}(\partial/\partial y^i)\delta/\delta j^1(y))\cdot$$

$$(\delta_{2j} + M_C{}^{-2}(\partial/\partial y^j)\, \delta/\delta j^2(y))\cdot$$

$$(\delta_{3k} + M_C{}^{-2}(\partial/\partial y^k)\, \delta/\delta j^3(y))Z_Y(j_\mu)\,\}\Big|_{j_\mu = 0} \qquad (8.30a)$$

After applying the functional derivatives the terms of order $M_C{}^{-2}$ sum to zero due to the gauge of the Y field and also due to setting the external source $j_\mu = 0$. The terms of order $M_C{}^{-6}$ are also set to zero when $j_\mu = 0$ since they each have a j_μ factor. Many terms of order $M_C{}^{-4}$ are set to zero when $j_\mu = 0$. The remaining terms of order $M_C{}^{-4}$ sum to zero. They are of the form:

$$\lim_{y_a \to y_b} [(\partial^2/\partial y_a{}^1 \partial y_b{}^2)(D_{F12}(y_a - y_b) - D_{F21}(y_a - y_b)) +$$

$$+ (\partial^2/\partial y_a{}^1 \partial y_b{}^3)(D_{F13}(y_a - y_b) - D_{F31}(y_a - y_b)) +$$

$$+ (\partial^2/\partial y_a{}^2 \partial y_b{}^3)(D_{F23}(y_a - y_b) - D_{F32}(y_a - y_b))] \qquad = 0$$

and thus equal zero since D_{Fij} is symmetric in i and j. Thus, effectively

$$\mathcal{J}(-i\delta/\delta j^\nu(y),y) \equiv 1 \qquad (8.30b)$$

This result is consistent with our physical notion that the integral does not generate dynamical effects. *Otherwise* a c-number interaction Lagrangian term would generate a non-trivial interacting quantum field theory effects – contrary to our physical expectations. Eq. 8.30b eliminates this physically unreasonable possibility.

Thus eq. 8.29 becomes

$$Z(J) = N \exp\{ i\int d^4y\, \mathscr{L}_{Fint}(-i\delta/\delta J_1(y), \ldots, -i\delta/\delta J_n(y))\}\cdot$$

$$\exp\{ -i\int d^4y_1 d^4y_2 \sum_k J_k(y_1)\Delta_{Fk}{}^{TT}(y_1 - y_2)J_k(y_2)/2]\} \qquad (8.31)$$

which gives the same perturbation theory found earlier using canonical quantum field theory that is built on the U matrix expansion.

Eq. 8.31 is based on the assumption that derivative couplings do not appear in the interaction Lagrangian. If the interaction Lagrangian does contain derivatives of scalar fields with respect to the X variable then the interaction Lagrangian has a superficial dependence on the functional derivative with respect to j^μ, which we can symbolize in the modified path integral expression:

$$Z(J) = N \{\exp\{ i\int d^4y\, \mathscr{L}_{Fint}(\partial/\partial(y^\nu + M_c^{-2}\delta/\delta j_\nu(y)), -i\delta/\delta J_1(y), \ldots, -i\delta/\delta J_n(y))\}\cdot$$

$$\exp\{-\sum_k [\, i\int d^4y_1 d^4y_2\, J_k(y_1)\Delta_{Fk}{}^{TT}(y_1 - y_2)J_k(y_2)/2]\}\, Z_Y(j_\mu)\, \}|_{j_\mu = 0} \qquad (8.32)$$

obtained from eq. 8.31. The j^μ dependence now appears only in the interaction Lagrangian. For good reason we will now excise the j^μ dependence from the interaction Lagrangian. We note that

$$i\Delta_{Fk}{}^{TT}(y_1 - y_2) = \langle 0|T(\phi(X(y_1)),\phi(X(y_2)))|0\rangle \qquad (6.51)$$

$$= i \int \frac{d^4p\, e^{-ip\cdot(y_1 - y_2)}\, R(\mathbf{p}, y_1 - y_2)}{(2\pi)^4\, (p^2 - m_k^2 + i\varepsilon)} \qquad (6.52)$$

with

$$R(\mathbf{p}, y_1 - y_2) = \exp[\, - p^i p^j \Delta_{Tij}(y_1 - y_2)/M_c^4] \qquad (6.53)$$

(summations are over space indices only in the Y Coulomb gauge) and

$$\Delta_{Tij}(z) = \int d^3k\ e^{-ik\bullet z}\ (\delta_{ij} - k_ik_j/\mathbf{k}^2)/[(2\pi)^3 2\omega_k] \tag{6.54}$$

We now define a more general two-tier propagator as:

$$i\Delta_{Fk}{}^{TT}(y_1 - y_2, z) \quad = \quad i \int \frac{d^4p\ e^{-ip\bullet(y_1 - y_2)}\ R(\mathbf{p}, z)}{(2\pi)^4\ (p^2 - m_k{}^2 + i\varepsilon)} \tag{8.33}$$

We then note that

$$\partial i\Delta_{Fk}{}^{TT}(y_1 - y_2)\ /\partial X^\mu = \partial i\Delta_{Fk}{}^{TT}(y_1 - y_2,\ z)/\partial y_1{}^\mu\ |_{z=y_1 - y_2} \tag{8.34}$$

comparing eqs. 6.51 and 6.52 with eq. 8.33. As a result we can write eq. 8.32 symbolically as:

$$Z(J) = N\ \{\exp\{\ i\int d^4y\ \mathscr{L}_{Fint}(\partial/\partial y^\nu,\ -i\delta/\delta J_1(y),\ \ldots,\ -i\delta/\delta J_n(y))\}\cdot$$

$$\exp\{-\sum_k [\ i\int d^4y_1 d^4y_2\ J_k(y_1)\Delta_{Fk}{}^{TT}(y_1 - y_2,\ z)J_k(y_2)/2]\}\ Z_Y(j_\mu)\ \}|_{j_\mu = 0,\ z=y_1 - y_2} \tag{8.35}$$

which simplifies to:

$$Z(J) = N\ \{\exp\{\ i\int d^4y\ \mathscr{L}_{Fint}(\partial/\partial y^\nu,\ -i\delta/\delta J_1(y),\ \ldots,\ -i\delta/\delta J_n(y))\}\cdot$$

$$\exp\{-\sum_k [\ i\int d^4y_1 d^4y_2\ J_k(y_1)\Delta_{Fk}{}^{TT}(y_1 - y_2,\ z)J_k(y_2)/2]\}\}|\ _{z=y_1 - y_2} \tag{8.36}$$

where z is set equal to the current $y_1 - y_2$ variables in each propagator, term by term, in the perturbative expansion of eq. 8.36 after performing any derivatives with respect to y (appearing in the interaction Lagrangian) on the propagator. We thus achieve a path integral expression that is very similar to the corresponding expression in conventional field theory – the only difference is the form of the free field propagators, which now contain a Gaussian factor.

ϕ^4 Theory Path Integral Formulation Example

We will now consider the case of two-tier scalar field theory using the specific example of the ϕ^4 Lagrangian of eqs. 5.20 – 5.24:

$$\mathscr{L}_{\text{Fint}}(X^\mu) = \tfrac{1}{4!}\, \chi_0\, \phi(X)^4 + \tfrac{1}{2}\, (m^2 - m_0{}^2)\phi(X)^2 \qquad (5.20)$$

with

$$\mathscr{L}_{\text{F0}} = \tfrac{1}{2}\, [\, (\partial\phi/\partial X^\nu)^2 - m^2\phi^2] \qquad (8.37)$$

In this theory eq. 8.31 can be written as:

$$Z(J) = N \exp\{\, i\textstyle\int d^4y\, \mathscr{L}_{\text{Fint}}(-i\delta/\delta J(y))\} \exp\{-i\textstyle\int d^4y_1 d^4y_2\, J(y_1)\Delta_{\text{Fk}}{}^{\text{TT}}(y_1 - y_2)J(y_2)/2\} \qquad (8.38)$$

The perturbation theory generated from eq. 8.38 is the same as conventional perturbation theory except for the differing free field propagators.

Two-Tier Yang-Mills Gauge Fields

Two-tier Yang-Mills gauge field theories have many similarities to conventional Yang-Mills theories.[35] We assume the reader in familiar with internal symmetries and the conventional Yang-Mills formulation.

General Rule: All gauge fields and derivatives of gauge fields, as well as group properties and other features such as the Fadeev-Popov method, are expressed solely in terms of the X coordinate system (which in turn is a function of the y coordinates).

We note all matter fields are assumed to be functions of X coordinates only as done in previous chapters. The general rule is implemented by defining:

1. The covariant derivative of any matter field $\Psi(X)$ by

$$D^\mu\Psi(X) = [\partial/\partial X_\mu + igA^\mu(X)]\Psi(X) \qquad (8.39)$$

with $A^\mu(X)$ being an element of a Lie algebra:

$$A^\mu(X) = A_a{}^\mu(X)L_a \qquad (8.40)$$

where L_a is a generator of a Lie algebra (a and b are internal symmetry indexes) with commutation relations:

[35] C. N. Yang and R. L. Mills, Phys. Rev. **96**, 191 (1954).

$$[L_a, L_b] = ic_{ab}{}^{c}L_c \tag{8.41}$$

with $c_{ab}{}^{c}$ being real numbers called the *structure constants* of the Lie algebra.

2. The field strengths are defined as the commutator of covariant derivatives:

$$F^{\mu\nu} = [D^\mu, D^\nu] = \partial A^\nu(X)/\partial X_\mu - \partial A^\mu(X)/\partial X_\nu + ig[A^\mu(X), A^\nu(X)] \tag{8.42}$$

3. The Lagrangian for a Yang-Mills gauge field interacting with a matter field $\psi(X)$ has the form:

$$\mathscr{L}_{YM} = [\mathscr{L}_F{}^{YM}(X^\mu) + \mathscr{L}_F{}^{Matter}(X^\mu)]\, J + \mathscr{L}_C(X^\mu(y), \partial X^\mu(y)/\partial y^\nu) \tag{8.43}$$

where J is the Jacobian and

$$\mathscr{L}_F{}^{YM}(X^\mu) = -\tfrac{1}{4}\, F_a{}^{\mu\nu}\, F_{a\mu\nu} \tag{8.44}$$

with $F^{\mu\nu} = F_a{}^{\mu\nu}L_a$ and

$$\mathscr{L}_F{}^{Matter}(X^\mu) = \mathscr{L}_F{}^{Matter}(\psi(X), D^\mu\psi(X)) \tag{8.45}$$

and $\mathscr{L}_C$ is defined as previously and specifies the Y field evolution.

The generalization to multiple gauge fields interacting with multiple matter fields is direct. The overall form of the Standard Model Lagrangian is:

$$L = \int d^4y\, \{J[\mathscr{L}_F{}^{Matter}(X^\mu) + \mathscr{L}_F{}^{GaugeFields}(X^\mu) + \mathscr{L}_F{}^{Higgs}(X^\mu)] + \\ + \mathscr{L}_C(X^\mu(y), \partial X^\mu(y)/\partial y^\nu)\} \tag{8.46}$$

where J is the Jacobian for the transformation from y to X coordinate integration. Eq. 8.46 can be rewritten as

$$L = \int d^4X[\mathscr{L}_F{}^{Matter}(X^\mu) + \mathscr{L}_F{}^{GaugeFields}(X^\mu) + \mathscr{L}_F{}^{Higgs}(X^\mu)] + \\ + \int d^4y\, \mathscr{L}_C(X^\mu(y), \partial X^\mu(y)/\partial y^\nu) \tag{8.47}$$

It is clear from eq. 8.47 that the conventional Standard Model equations of motion and canonical quantization procedure emerge in the two-tier formulation if all expressions are written as functions of the X coordinates as shown previously in our discussions of separable Lagrangians. The second quantization of X as a function of the y coordinates leads to the gaussian factor in all free particle propagators (except the Y propagator).

Thus the gauge field and matter field parts of the Lagrangian, the gauge field transformation laws and other related operations solely depend on the X coordinate system as stated in the general rule above. The quantization and field equations are the same as the conventional case – except that they are specified solely in terms of the X coordinate system.

Path Integral Formulation and Fadeev-Popov Method

We now turn to the two-tier path integral formulation of Yang-Mills gauge theories and in particular to the two-tier version of the Fadeev-Popov method. The two-tier path integral for a gauge field can be written symbolically as:

$$Z(J^\mu) = N \int DADY\Delta_{FP}(A)\delta(F(A)) \exp\{i\int d^4y\, [\mathscr{L} + j_\mu(y)Y^\mu(y)+J^\mu(y)A_\mu(X)]\}|_{j_\mu = 0} \qquad (8.48)$$

where $\delta(F(A))$ specifies the gauge and $\Delta_{FP}(A)$ is the Fadeev-Popov determinant. The Lagrangian is

$$\mathscr{L} = J\mathscr{L}_F{}^{YM}(X^\mu) + \mathscr{L}_C(X^\mu(y), \partial X^\mu(y)/\partial y^\nu) \qquad (8.49)$$

with $\mathscr{L}_F{}^{YM}$ specified by eq. 8.44 and J the Jacobian for the transformation from y coordinates to X coordinates. The Fadeev-Popov determinant may be calculated in the standard way. First we note that the delta function fixing the gauge can be written as a delta function in the gauge times a determinant:

$$\delta(F(A^\omega)) = \delta(\omega - \omega_0)\, \left|\det \delta F(A_\mu{}^\omega(X))/\delta\omega(X)\right|^{-1}\Big|_{F(A)=0} \qquad (8.50)$$

where ω_0 is a reference gauge, where

$$A_\mu{}^\omega(X) = A_\mu(X) + \partial\omega(X)/\partial X^\mu \qquad (8.51)$$

and where

$$\Delta_{FP}(A) = \left|\det \delta F(A_\mu{}^\omega(X))/\delta\omega(X)\right|\Big|_{F(A)=0} \qquad (8.52)$$

We will choose the Lorentz gauge to evaluate the Fadeev-Popov determinant:

$$F_a(A) = \partial A_a^{\mu}(X)/\partial X^{\mu} = 0 \tag{8.53}$$

Under an infinitesimal gauge transformation:

$$A_{a_\mu}{}^{\omega}(X) = A_{a_\mu}(X) + g^{-1}\partial\omega_a/\partial X^{\mu} + c_{ab}{}^{c}\,\omega_b(X)A_c^{\mu}(X) \tag{8.54}$$

we find

$$F_a(A_\mu{}^{\omega}(X)) = \partial\,(A_{a_\mu}(X) + g^{-1}\partial\,\omega_a(X)/\partial X^{\mu} + c_{ab}{}^{c}\,\omega_b(X)A_c^{\mu}(X))/\partial X^{\mu}$$

$$= g^{-1}\,\Box_X\,\omega_a(X) + c_{ab}{}^{c}\,\partial\,\omega_b(X)/\partial X^{\mu}\,A_c^{\mu}(X) \tag{8.55}$$

Thus

$$\delta F_a(A_\mu{}^{\omega}(X))/\delta\omega_b(X) = g^{-1}\,\delta_{ab}\Box_X + c_{ab}{}^{c}A_c^{\mu}(X)\partial/\partial X^{\mu} \tag{8.56}$$

and

$$\Delta_{FP}(A) = \left|\det\,(g^{-1}\,\delta_{ab}\Box_X + c_{ab}{}^{c}A_c^{\mu}(X)\partial/\partial X^{\mu})\right|\Big|_{F(A)=0} \tag{8.57}$$

We note the two-tier Fadeev-Popov determinant is solely a function of the X coordinates. Thus we can follow the standard procedure and rewrite the determinant as a path integral over anti-commuting c-number fields with a ghost Lagrangian that is solely a function of X – just like the gauge particles and other particles in two-tier theories:

$$\Delta_{FP}(A) = \int D\chi^{*}D\chi\,\exp[\,i\int d^4X\,\mathscr{L}^{ghost}(X^{\mu})] \tag{8.58}$$

where

$$\mathscr{L}^{ghost}(X^{\mu}) = \chi_a{}^{*}(X)[\delta_{ab}\Box_X + g\,c_{ab}{}^{c}A_c^{\mu}(X)\partial/\partial X^{\mu}]\chi_b(X) \tag{8.59}$$

Thus the complete path integral is

$$Z(J^\mu) = N \int DAD\chi^* D\chi DY \delta(F(A)) \exp\{i\int d^4y\, [\mathscr{L} +$$

$$j_\mu(y)Y^\mu(y)+J^\mu(y)A_\mu(X)]\}|_{j_\mu = 0} \qquad (8.60)$$

where $\delta(F(A))$ specifies the gauge and

$$\mathscr{L} = J\mathscr{L}_F{}^{YM}(X^\mu) + J\mathscr{L}^{ghost}(X^\mu) + \mathscr{L}_C(X^\mu(y), \partial X^\mu(y)/\partial y^\nu) \qquad (8.61)$$

with $\mathscr{L}_F{}^{YM}$ specified by eq. 8.44.

At this point it is obvious that we can follow almost identical steps as we did in the scalar particle case starting from eq. 8.9 and obtain an expression similar to eq. 8.38. The result is a perturbation theory for the Yang-Mills gauge field that is identical to the usual theory except that the free propagators for the gauge fields, and the ghost fields, acquire the gaussian factor R(p,z) as stated earlier in the discussions of eqs. 6.52, 6.88, 6.95, 6.102, and 7.120-123.

Two-Tier Massive Vector Fields

Massive vector fields have been a problem for perturbation theory due to the $k_\mu k_\nu/m^2$ term appearing in the free field propagator. This term makes conventional interacting quantum field theories of massive vector bosons non-renormalizable. The Higgs mechanism is used in Electroweak theory to evade the non-renormalizability of massive gauge fields. It gives mass to the vector bosons mediating the weak force while maintaining the renormalizability of the theory.

In Two-Tier quantum field theory a massive vector boson does not create renormalization issues. For example the two-tier version of the Weinberg-Salam model of the electromagnetic and weak forces (with massive vector bosons) is finite! Thus the need for spontaneous symmetry breaking to give mass to vector bosons in Electroweak theory is not present. Higgs particles are not needed (although they may exist. Their existence is an experimental question – not a theoretical necessity for a two-tier Electroweak theory with massive vector bosons.)

Massive Vector Boson Propagator

The massive free vector particle propagator in *conventional* quantum field theory has the representation:

$$i\Delta_{FV}(y_1 - y_2)_{\mu\nu} = -i \int \frac{d^4k\, e^{-ik\bullet(y_1 - y_2)} (g_{\mu\nu} - k_\mu k_\nu/m^2)}{(2\pi)^4 (k^2 - m^2 + i\varepsilon)} \tag{8.62}$$

A two-tier massive vector boson theory can be constructed in a straightforward way from the Lagrangian:

$$\mathscr{L} = [-¼\, F_V{}^{\mu\nu}(X(y))F_{V_{\mu\nu}}(X(y)) - ½\, m^2\, V^\mu V_\mu]J + \mathscr{L}_C(X^\mu(y),\, \partial X^\mu(y)/\partial y^\nu, y) \tag{8.63}$$

with J the Jacobian and with

$$F_{V_{\mu\nu}}(X(y)) = \partial V_\mu(X(y))/\partial X^\nu - \partial V_\nu(X(y))\,/\partial X^\mu \tag{8.64}$$

and the usual two-tier Y Lagrangian terms

$$\mathscr{L}_C(X^\mu(y),\, \partial X^\mu(y)/\partial y^\nu,\, y) = -¼\, F_Y{}^{\mu\nu}F_{Y_{\mu\nu}} \tag{7.4}$$

$$F_{Y_{\mu\nu}} = (\partial Y_\mu/\partial y^\nu - \partial Y_\nu\,/\partial y^\mu) \tag{7.5}$$

Following steps similar to the previously considered scalar particle, fermion and massless vector particle (photon) cases we find the two-tier massive vector boson Feynman propagator to be:

$$i\Delta_{FV}{}^{TT}(y_1 - y_2)_{\mu\nu} = -i \int \frac{d^4k\, e^{-ik\bullet(y_1 - y_2)} (g_{\mu\nu} - k_\mu k_\nu/m^2)\, R(\mathbf{k},\, y_1 - y_2)}{(2\pi)^4 (k^2 - m^2 + i\varepsilon)} \tag{8.65}$$

No Need for Higgs Mechanism in Electroweak Theory

The leading coordinate space dependence at high energy (short distance) of the fourier transform of $\Delta_{FV}{}^{TT}$ is the same as $\Delta_F{}^{TT}$ since it comes from the $g_{\mu\nu}$ term in $\Delta_F{}^{TT}$

$$\Delta_{FV}{}^{TT}(y_1 - y_2)_{\mu\nu} \sim g_{\mu\nu}\, (y_1 - y_2)^2 \tag{8.66}$$

which in momentum space is equivalent to

$$\Delta_{FV}{}^{TT}(p)_{\mu\nu} \sim g_{\mu\nu}\, p^{-6} \tag{8.67}$$

The $k_\mu k_\nu/m^2$ term appearing in the free vector field two-tier propagator actually is of higher order in the large momentum limit (short distance):

$$\int \frac{d^4k\, e^{-ik\bullet(y_1 - y_2)} k_\mu k_\nu/m^2)\, R(\mathbf{k}, y_1 - y_2)}{(2\pi)^4\, (k^2 - m^2 + i\varepsilon)} \sim (y_1 - y_2)^4 \tag{8.68}$$

which corresponds to momentum space behavior of

$$p^{-8} \tag{8.69}$$

Two-Tier propagators such as the propagator in eq. 8.65 have the feature that the gaussian factor "inverts" the high-energy behavior of the terms in the numerator of the integrand: terms with higher powers of momentum are less significant at short distances. The term with the lowest power of momentum generates the leading behavior at high energy (short distances).

Thus two-tier massive vector boson theories are ultra-violet convergent and do not constitute a problem as they do in conventional quantum field theory. *Therefore there is no need for the Higgs mechanism in the Electroweak sector of the Standard Model in order to obtain a renormalizable theory. Ordinary massive vector bosons can be used and the resulting two-tier theory is finite!*

General Short Distance (High Momentum) Behavior of Two-Tier Propagator

The higher the power of the momentum in the numerator of the integrand of a two-tier propagator, the more convergent the large momentum behavior of the fourier transform of the two-tier propagator.

The short distance behavior of a term with n factors of momentum in a two-tier propagator has the leading short distance coordinate space behavior:

$$\int \frac{d^4k\, e^{-ik\bullet(y_1 - y_2)}\, k_{\mu 1} k_{\mu 2} \ldots k_{\mu n}\, R(\mathbf{k}, y_1 - y_2)}{(2\pi)^4\, (k^2 - m^2 + i\varepsilon)} \sim (y_1 - y_2)^{2+n} \tag{8.70}$$

which corresponds to the high energy behavior:

$$p^{-6-n} \tag{8.71}$$

In contrast to conventional quantum field theory the more powers of momentum in the numerator of a two-tier propagator, the better the short distance behavior!

Higgs Particles

A previous section shows that the Higgs Mechanism is not needed in the two-tier ElectroWeak sector of the Standard Model. Massive vector bosons such as **W**'s would not make the Electroweak theory non-renormalizable. *As we have seen a Two-tier Electroweak theory with massive vector bosons is finite.*

Nevertheless we would like to point out a two-tier version of the Higgs particle sector of the Standard Model can be defined that largely parallels the conventional treatment of the Higgs sector. Higgs particles continue to be of interest since they may play a role in the origin of particle masses.

The two-tier scalar Higgs field Lagrangian terms (plus $\mathscr{L}_C$) can be written:

$$\mathscr{L}_{Higgs} = J[D_\mu\phi^\dagger D^\mu\phi - V(\phi(X))] + \mathscr{L}_C \tag{8.72}$$

$$\begin{aligned} V(\phi(X)) = \ & m_0^2\phi^\dagger(X(y))\phi(X(y)) + \lambda[\phi^\dagger(X(y))\phi(X(y))]^2 + \\ & + G_c[\bar{L}(X)\phi(X)R(X) + \bar{R}(X)\phi^\dagger(X)R(X)] \end{aligned} \tag{8.73}$$

with the covariant derivative defined with the usual B_μ and $W_\mu{}^a$ gauge fields of the Standard Model and L representing a left-handed fermion isodoublet and R representing a right-handed fermion isosinglet. We note all items in eqs. 8.72 and 8.73 are written solely as functions of the X coordinates with the exception of $\mathscr{L}_C$ which is the Lagrangian term for the Y field.

The conventional effective potential method can be followed to implement the Higgs mechanism. In particular we may write

$$\phi(X(y)) = \langle\phi\rangle + \eta(X) \tag{8.74}$$

with $\langle\phi\rangle$ the vacuum expectation value, which is a constant by translational invariance. Then vector bosons can acquire mass via the Higgs mechanism. The quantum part of the Higgs field has a two-tier propagator with p^{-6} behavior for large momentum. Thus all sectors of the Standard Model wind up with two-tier propagators and all perturbative calculations are finite.

General Form of the Two-Tier Standard Model Path Integral

The general form of the path integral for the two-tier version of the Standard Model is:

$$Z(J) = N \{\exp\{ i\int d^4y \, \mathscr{L}_{Fint}(\partial/\partial y^\nu, -i\delta/\delta J_1(y), \ldots, -i\delta/\delta J_n(y))\} \cdot$$

$$\exp\{-\sum_k [\, i\int d^4y_1 d^4y_2 \, J_k(y_1)\Delta_{Fk}{}^{TT}(y_1 - y_2, z) J_k(y_2)/2]\}\}\Big|_{z=y_1-y_2} \qquad (8.75)$$

where the sum over k is a sum over all matter fields, gauge fields, Higgs fields, and ghost fields. The index k, and indices on the functional derivatives, represent all space-time indices and internal symmetry indices that are relevant for each particle. We assume the total number of particle and ghost fields is n. Notice all dependence on the Y field has been "integrated away."

Also derivatives with respect to y appearing in the interaction Lagrangian are applied to two-tier propagators using the method represented by eqs. 8.33 and 8.34. This procedure resullts in the same momentum polynomials for Feynman diagrams in a two-tier version as appear in the corresponding conventional theory. Thus we have the same algebraic structure (both in the momentum polynomials and internal symmetries) in the two-tier version of a conventional theory.

We note that the large distance behavior of the two-tier theory is the same as the Standard Model in all respects including gauge symmetries. Deviations from the conventional Standard Model results only appear at extremely high energies of the order of M_c.

Renormalization - Finite

Since all particle (and ghost) propagators are two-tier propagators in the two-tier Standard Model, the two-tier Standard Model yields finite results to all orders in perturbation theory. We note the large momentum behavior of the various types of particle propagators in the two-tier Standard Model is:

<u>Fermion Propagators</u>

$$p^{-7}$$

<u>Vector Boson (Gauge Field) Propagators</u>

$$p^{-6}$$

<u>Ghost Propagators</u>

$$p^{-6}$$

Higgs Particle Propagators

$$p^{-6}$$

Thus all Feynman diagrams are highly ultra-violet convergent and the two-tier Standard Model is finite. This result is independent of the details of the internal symmetries, particle spectrum, and particle masses.

Unitarity

The two-tier Standard Model *superficially* appears to have a unitarity problem due to the non-hermitean nature of its hamiltonian. The lack of hermiticity is due entirely to the appearance of iY^μ in the X^μ field coordinates.

Thus the two-tier Standard Model interaction hamiltonian is not hermitean:

$$H_{Fint} = \int d^3y' \mathscr{H}_{Fint}(y' + iY(y')/M_c^2) \tag{8.76}$$

but

$$H_{Fint}^\dagger = \int d^3y' \mathscr{H}_{Fint}(y' - iY(y')/M_c^2) \neq H_{Fint} \tag{8.77}$$

The relation between H_{Fint} and its hermitean conjugate is

$$H_{Fint} = V\, H_{Fint}^\dagger\, V \tag{8.78}$$

where $V^2 = I$ is the metric operator defined in eqs. 5.16 – 5.18. Eq. 8.78 implies that the two-tier Standard Model S matrix is not unitary. The two-tier Standard Model S matrix is pseudo-unitary:

$$S^{-1} = V\, S^\dagger\, V \tag{8.79}$$

Therefore

$$S^\dagger V S = V \tag{8.80}$$

We will now show that *the Two-Tier Standard Model S matrix is effectively unitary between physical asymptotic states* – states consisting of only physical particles: leptons, quarks, photons, W and Z particles, gluons, and Higgs particles (if they exist). Put another way: states of all particles in the two-tier Standard Model except ghosts and Y particles.

As we noted in an earlier discussion of coherent states, the S matrix is *block diagonal* in the sense that in-states consisting of only physical particles will only produce out-states consisting of physical particles. The consequence of block diagonality is that *S is unitary within the physical states sector.* To prove this point consider an S matrix element between physical asymptotic states:

$$\langle f\ in|V|i\ in\rangle = \langle f\ in|S^\dagger VS|i\ in\rangle = \sum_{n,m} \langle f\ in|S^\dagger|n\rangle\langle n|V|m\rangle\langle m|S|i\ in\rangle \tag{8.81}$$

where the block diagonality of S (and the diagonality of V) limits the intermediate states |n> and |m> to states containing physical particles only. We note V is diagonal with eigenvalue one between states consisting of physical particles only. Thus eq. 8.81 becomes

$$\delta_{fi} = \sum_n \langle f\ in|S^\dagger|n\rangle\langle n|S|i\ in\rangle \tag{8.82}$$

or

$$S_{phys}^\dagger S_{phys} = I$$

with the result

$$S_{phys}^\dagger = S_{phys}^{-1} \tag{8.83}$$

thus proving unitarity for physical states.

Anomalies

The axial anomaly (Adler-Bell-Jackiw anomaly) follows from the linear divergence of a fermion triangle graph (Fig. 8.1) in the conventional Standard Model.

In two-tier theory the axial anomaly does not appear to be present. The two-tier formalism parallels the conventional approach. A massless Dirac field theory formally is invariant under a chiral transformation and a conserved axial-vector current formally results. The two-tier axial-vector current is

$$j_5^\mu(X(y)) = \bar{\psi}(X(y))\gamma^\mu\gamma_5\psi(X(y)) \tag{8.84}$$

with formal conservation law:

$$\partial j_5^\mu(X(y))/\partial X^\mu = 2m\ j_5(X(y)) = 2m\ \bar{\psi}(X(y))\gamma_5\psi(X(y)) \tag{8.85}$$

(Note the conventional formalism has the same form except that the X coordinates are purely coordinates and not fields.) Eq. 8.85 gives

$$\partial j_5^{\mu}(X(y))/\partial X^{\mu} = 0 \tag{8.86}$$

in the limit m → 0. The question we now address is whether eq. 8.86 holds in two-tier perturbation theory – perhaps in the same form as the conventional axial anomaly:

$$\partial j_5^{\mu}(X(y))/\partial X^{\mu} = 2m\, j_5(X(y)) + \alpha_0(4\pi)^{-1}\varepsilon^{\mu\nu\alpha\beta}F_{\alpha\beta}F_{\mu\nu} \quad ? \tag{8.87}$$

where α_0 is the unrenormalized fine structure constant.

The simplest manifestation of the axial anomaly in conventional field theory is the fermion triangle diagram, which we will now examine in two-tier quantum field theory. Unlike the conventional quantum field theory result, the two-tier triangle diagram is finite and zero. *Thus the axial anomaly does not appear to exist in two-tier quantum field theory.*

The absence of the anomaly reflects the self-regularization of two-tier quantum field theory, which preserves chiral invariance. Unlike Pauli-Villars regularization, for example, the regularization in two-tier theory does not use a massive fermion. Unlike the dimensional regularization approach (where there is no equivalent to γ_5), two-tier theory can use the normal γ_5 matrix.

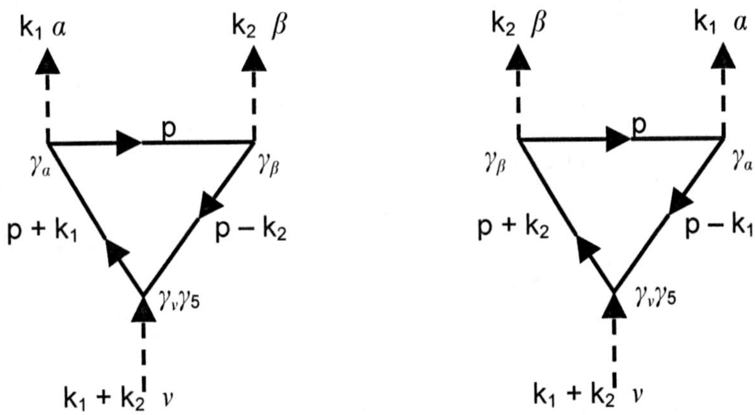

Figure 8.1. The V-V-A triangle diagrams.

The expression for the two-tier triangle diagrams is:

$$T_{\alpha\beta\nu}(k_1, k_2) = S_{\alpha\beta\nu}(k_1, k_2) + S_{\beta\alpha\nu}(k_2, k_1) \tag{8.88}$$

where

$$S_{\alpha\beta\nu}(k_1, k_2)\delta^4(k_1 + k_2 - q) = -iN\int d^4y_1 d^4y_2 d^4y_3\, e^{ik_1\cdot y_1 + ik_2\cdot y_2 - iq\cdot y_3}$$
$$\mathrm{Tr}\{S_F{}^{TT}(y_1 - y_3)\gamma_\alpha S_F{}^{TT}(y_2 - y_1)\gamma_\beta S_F{}^{TT}(y_3 - y_2)\gamma_\nu\gamma_5\}/(2\pi)^4 \tag{8.89}$$

where N is a constant, and where $S_F{}^{TT}(z)$ is specified by eq. 7.45. We now define the fourier transform:

$$S_F{}^{TT}(z) = -i\int d^4p\, e^{-ip.z}\, \mathscr{S}^{TT}(p)/(2\pi)^4 \tag{8.90}$$

where $\mathscr{S}^{TT}(p)$ defined by eq. 7.93. We then substitute the fourier transform in eq. 8.89 and perform the coordinate integrations to obtain:

$$S_{\alpha\beta\nu}(k_1, k_2) = N\int d^4p\, \mathrm{Tr}\{\mathscr{S}^{TT}(p + k_1)\gamma_\alpha \mathscr{S}^{TT}(p)\gamma_\beta \mathscr{S}^{TT}(p - k_2)\gamma_\nu\gamma_5\}/(2\pi)^4$$

We note that

$$k_1{}^\alpha T_{\alpha\beta\nu}(k_1, k_2) \neq 0 \tag{8.91}$$

$$k_2{}^\beta T_{\alpha\beta\nu}(k_1, k_2) \neq 0 \tag{8.92}$$

$$(k_1 + k_2)^\nu T_{\alpha\beta\nu}(k_1, k_2) \neq 0 \tag{8.93}$$

in two-tier quantum field theory because the conservation laws are expressed with respect to the X coordinates – not the y coordinates. Thus since $k_1{}^\alpha$ corresponds $\partial/\partial y_\alpha$, and not $\partial/\partial X_\alpha$ there is no reason for eqs. 8.91-93 to be zero. However at "large distances" relative to $M_c{}^{-1}$ we see

$$k_1{}^\alpha T_{\alpha\beta\nu}(k_1, k_2) \cong 0 \tag{8.94}$$

$$k_2{}^\beta T_{\alpha\beta\nu}(k_1, k_2) \cong 0 \tag{8.95}$$

$$(k_1 + k_2)^\nu T_{\alpha\beta\nu}(k_1, k_2) \cong 0 \tag{8.96}$$

to very good approximation since the gaussian damping factor in the fermion propagators is approximately unity and thus the two-tier expression becomes essentially the same as the conventional field theory expression.

On the other hand at very short distances the anomaly appears to be absent since two-tier theory is very well behaved at high energy with

$$\mathscr{S}^{TT}(p) \backsim \gamma^0 M_c^{\,6}\, p^{-7} + \mathscr{O}(p^{-9}) \tag{7.102}$$

As a result we see

$$k_1{}^{\alpha} T_{\alpha\beta\nu}(k_1, k_2) \sim p^{4-21} \sim p^{-17} \tag{8.97}$$

as $p \to \infty$ is highly convergent and thus there is no high energy divergence unlike conventional field theory where the integral is linearly divergent.

Asymptotic Freedom and Quark Confinement

Two-tier quantum field theory is totally consistent with the Standard Model at currently accessible energies. Thus if the Standard Model Quantum Chromodynamics (QCD) sector is asymptotically free, and also gives quark confinement at large distance (compared to M_c^{-1}), then similar statements would also be true in the two-tier version of QCD.

We assume M_c is extremely large – much beyond current energies – and possibly of the order of the Planck mass. Fig. 8.2 depicts the various regions in two-tier QCD assuming a very large M_c.

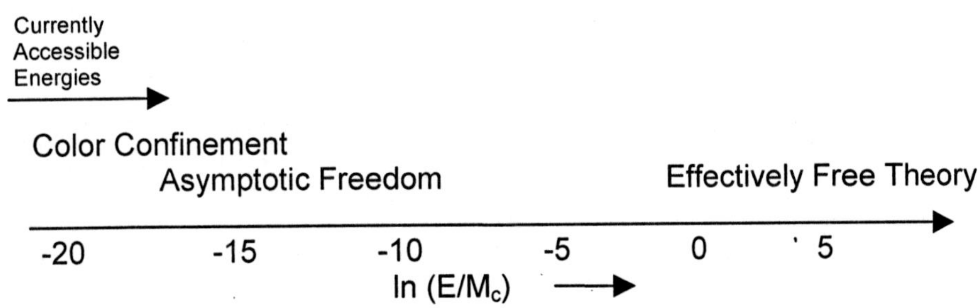

Figure 8.2. A depiction of the two-tier QCD regions as a function of the logarithm of the energy assuming M_c is of the order of the Planck mass.

Two-Tier Coordinates

We have developed a physical picture of two-tier coordinate systems in which we define two sets of related coordinates. This picture views ordinary real 4-dimensional space as a low energy approximation to a complex 4-dimensional space that only becomes apparent at ultra-high energies with the scale set by M_c. The imaginary part of the coordinates is based on the excitations of a quantum field Y^μ. The real part of the coordinates is the familiar 4-dimensional c-number coordinate system.

Another way to view the two-tier X and y coordinate systems is to view them as defining a four-dimensional hyperplane in an eight dimensional real space-time. At the present level of discussion this alternate view does not generate new insights.

It would also be interesting to consider an extension of the theory to the case where both the real and imaginary parts of the coordinate system are quantum fields. In this case the c-number coordinates of daily experience would emerge as a "condensation", or spontaneous symmetry breaking, phenomena.

One could also generalize the Y quantum field to a non-abelian gauge field, which, properly handled, could be the origin of internal symmetries such as QCD SU(3) using a Kaluza-Klein approach within a general relativistic *vierbein* framework. This approach is an alternative to the compactification of dimensions that is a cornerstone of SuperString theories.

9. Two-Tier Quantum Gravity: Finite

Introduction

There are numerous excellent books and monographs on classical gravity and a large literature on quantum gravity.[36] Therefore our discussion will assume the reader is familiar with classical General Relativity and aware of attempts to create quantum theories of gravity.

We will begin by establishing the general form of two-tier classical General Relativity and then proceed to define a quantization procedure. We work in Minkowski space with three space and one time dimension. The flat-space metric $\eta_{\alpha\beta}$ is defined as diagonal with $\eta_{00} = 1$ and $\eta_{ij} = -\delta_{ij}$ for i, j = 1,2,3.

Two-Tier General Relativity

In developing Quantum Gravity we will make the same ansatz that we have made throughout our development of two-tier quantum field theories: all field expressions are functions of the X coordinate field system, which in turn are functions of the "ordinary" y space-time coordinate system. The y coordinates can undergo general relativistic transformations. The X coordinate system thereby undergoes General Relativistic transformations appropriate to vectors.

---

[36] H. Weyl, *Space, Time, Matter* (Dover, New York, 1950); L. D. Landau and E. M. Lifshitz, *The Classical Theory of Fields*, (Addison-Wesley, New York, 1962); S. Weinberg, *Gravitation and Cosmology*, (John Wiley & Sons, New York, 1972); C. W. Misner, K. S. Thorne and J. A. Wheeler, *Gravitation*, (W. H. Freeman, San Francisco, 1973); B. S. DeWitt, Phys. Rev. **162**, 1239 (1967), **162**, B1195 (1967); R. P. Feynman, Acta Physica Polonica **24**, 697 (1963); S. Deser and P. van Nieuwenhuizen, Phys. Rev. Letters **32**, 245 (1974); S. Deser, H.-S. Tsao and P. van Nieuwenhuizen, "One Loop Divergences of the Einstein-Yang-Mills System", Brandeis Univ. preprint (1974); S. Weinberg, Phys. Rev. **138**, B988 (1965); L. Smolin, *Three Roads to Quantum Gravity*, (Basic Books, New York, 2001); L. Smolin, "How Far are We From the Quantum Theory of Gravity", (Univ. Waterloo preprint (2003) and references therein; T. Thiemann, "Lectures on Loop Quantum Gravity", Preprint AEI-2002-087, Albert Einstein Insitute, Golm, Germany (2002) and references therein; A. pais and G. E. Uhlenbeck, Phys. Rev. **79**, 145 (1950); G. E. Uhlenbeck, "Lecture Notes on General Relativity", The Rockefeller University (1967), unpublished; S. Blaha, "Generalization of Weyl's Unified Theory to Encompass a Non-Abelian Internal Symmetry Group" SLAC-PUB-1799, Aug 1976; S. Blaha, "Quantum Gravity and Quark Confinement" Lett. Nuovo Cim. **18**, 60 (1977); R. Utiyama, Phys. Rev. **101**, 1597 (1956); T. W. B. Kibble, J. Math. Phys. **2**, 212 (1961); R. Arnowitt, S. Deser, and C. W. Misner, Phys. Rev. **117**, 1595 (1960); and references therein.

We will define the proper time differential $d\tau$ as

$$d\tau^2 = g_{\mu\nu}(X(y))dX^\mu dX^\nu \tag{9.1}$$

with

$$X^\mu(y) = y^\mu + i\, Y^\mu(y)/M_c^2 \tag{3.12}$$

Thus eq. 9.1 could be written:

$$d\tau^2 = g_{\mu\nu}(X(y))(\eta^\mu{}_\alpha + iM_c^{-2}\partial Y^\mu/\partial y^\alpha)(\eta^\nu{}_\beta + iM_c^{-2}\partial Y^\nu/\partial y^\beta)dy^\alpha dy^\beta \tag{9.2}$$

The inverse of $g_{\mu\nu}$, denoted $g^{\nu\lambda}$, satisfies

$$g_{\mu\nu}(X(y))g^{\nu\lambda}(X(y)) = \delta_\mu{}^\lambda \tag{9.3}$$

Since the algebraic manipulation of the tensor indices is the same as that of the conventional theory of gravitation we have the two-tier affine connection expression:

$${}_X\Gamma^\sigma{}_{\lambda\mu} = \tfrac{1}{2}\, g^{\nu\sigma}\{\partial g_{\mu\nu}/\partial X^\lambda + \partial g_{\lambda\nu}/\partial X^\mu - \partial g_{\lambda\mu}/\partial X^\nu\} \tag{9.4}$$

The two-tier Riemann-Christoffel curvature tensor is:

$${}_X R^\lambda{}_{\mu\nu\kappa} \equiv \partial_X\Gamma^\lambda{}_{\mu\nu}/\partial X^\kappa - \partial_X\Gamma^\lambda{}_{\mu\kappa}/\partial X^\nu + {}_X\Gamma^\alpha{}_{\mu\nu}\, {}_X\Gamma^\lambda{}_{\kappa\alpha} - {}_X\Gamma^\alpha{}_{\mu\kappa}\, {}_X\Gamma^\lambda{}_{\nu\alpha} \tag{9.5}$$

and the two-tier Ricci tensor is

$${}_X R_{\mu\nu} = {}_X R^\alpha{}_{\mu\alpha\nu} \tag{9.6}$$

The two-tier curvature scalar is

$${}_X R = g^{\mu\nu}\, {}_X R_{\mu\nu} \tag{9.7}$$

We also define

$${}_X R_{\lambda\mu\nu\kappa} = g_{\lambda\alpha}\, {}_X R^\alpha{}_{\mu\nu\kappa} \tag{9.8}$$

with the result

$$_XR_{\lambda\mu\nu\kappa} = ½\,[\partial^2 g_{\lambda\nu}/\partial X^\kappa\partial X^\mu - \partial^2 g_{\mu\nu}/\partial X^\kappa\partial X^\lambda - \partial^2 g_{\lambda\kappa}/\partial X^\nu\partial X^\mu + \partial^2 g_{\mu\kappa}/\partial X^\nu\partial X^\lambda]$$

$$+ g_{\alpha\beta}[\,_X\Gamma^\alpha{}_{\nu\lambda}\,{}_X\Gamma^\beta{}_{\mu\kappa} - {}_X\Gamma^\alpha{}_{\kappa\lambda}\,{}_X\Gamma^\beta{}_{\mu\nu}] \tag{9.9}$$

The algebraic properties, and the Bianchi identities, satisfied by $_XR_{\lambda\mu\nu\kappa}$ in the two-tier theory of gravitation are identical to those of the conventional theory with all derivatives being with respect to X.

The two-tier version of Einstein's field equations is:

$$_XR_{\mu\nu} - ½\, g_{\mu\nu}\,{}_XR = -8\pi G\, T_{\mu\nu} \tag{9.10}$$

where G is Newton's gravitational constant (6.673×10^{-11} $m^3kg^{-1}s^{-2}$) and $T_{\mu\nu}$ is the energy-momentum tensor. It is convenient to define the coupling constant

$$\kappa = \sqrt{4\pi G} \tag{9.11}$$

Lagrangian Formulation

We will now formulate a two-tier Quantum Gravity theory following the same ansatz that we have used throughout this book.

Unified Standard Model and Quantum Gravity Lagrangian

We define the Lagrangian, and action, for the unified quantum field theory of gravitation and the Standard Model as

$$L_{Unified} = \int d^4y\, \mathscr{L}_{Unified} \tag{9.12}$$

$$\mathscr{L}_{Unified} = J\sqrt{g(X)}\,\left(\mathscr{L}_F{}^{Grav}(X^\mu) + \mathscr{L}_F{}^{SM}(X^\mu)\right) + \mathscr{L}_C \tag{9.13}$$

with

$$\mathscr{L}_F{}^{Grav}(X^\mu) = (2\kappa^2)^{-1}{}_XR \tag{9.14}$$

where $\mathscr{L}_F{}^{SM}$ is the complete "normal" Quantum Field theory Lagrangian for the Standard Model variant under consideration written in a general covariant form, g(X) is the absolute value of the determinant of $g_{\mu\nu}$, and J is the Jacobian of eq. 2.21. *All particle fields in $\mathscr{L}_F{}^{SM}$ are assumed to be functions of the X^μ coordinate only. The dependence of the particle fields on the* *"underlying" coordinates y^μ is assumed to be solely through X^μ.* The Lagrangian $\mathscr{L}_{\text{Unified}}$ is a separable Lagrangian of the type of eq. 2.26 embodying the composition of extrema described in chapter 2.

As in all of these cases we have specified the coordinate part of the Lagrangian $\mathscr{L}_{\text{C}}$ as

$$\mathscr{L}_{\text{C}} = +\tfrac{1}{4}\, M_c^{\,4} F^{\mu\nu} F_{\mu\nu} \tag{3.10}$$

with

$$F_{\mu\nu} = \partial X_\mu/\partial y^\nu - \partial X_\nu/\partial y^\mu \tag{3.11}$$

$$X_\mu(y) = y_\mu + i\, Y_\mu(y)/M_c^{\,2} \tag{3.12}$$

This Lagrangian is re-expressed as

$$\mathscr{L}_{\text{C}} = -\tfrac{1}{4}\, F_Y^{\mu\nu} F_{Y\mu\nu} \tag{3.15}$$

with

$$F_{Y\mu\nu} = \partial Y_\mu/\partial y^\nu - \partial Y_\nu/\partial y^\mu \tag{3.14}$$

and

$$F_Y^{\mu\nu} = \eta^{\mu\alpha}\eta^{\nu\beta} F_{Y\alpha\beta} \tag{9.15}$$

Why Are the Y Field Dynamics Independent of the Gravitational Field?

It is evident from eqs. 9.12-3 and 9.15 that the Y field is truly free and, in particular, does not depend on the gravitational field as represented by $\sqrt{g}$ and $g_{\mu\nu}$. Our rationale for this formulation is: the Y field is a quantum field at each point in space-time including regions with ultra-strong gravitational fields such as the neighborhoods of black holes. If Y were to depend on the gravitational field then the Y field could be appreciable in such regions and might even be a "classical" field. In this case we would have new dimensions, allbeit imaginary, for which no evidence exists.

Furthermore, the Y part of the Lagrangian establishes a functional relation between the imaginary Y coordinates and the real space-time y coordinates. The Principle of Equivalence applies only to the real coordinates and does not apply to imaginary Quantum Dimensions™. Thus there is no reason to require the Y part of the action to be invariant under general coordinate transformations.

Lastly, it might be argued that the non-invariance of the Y part of the action under general coordinate transformations effectively creates an "absolute" coordinate system. The local quantum nature of the Y field serves to evade this possibility. At any non-singular space-time point we can establish a locally inertial coordinate system in which matter conforms to the laws of special relativity. The Y quantum field may be defined in that coordinate system and the dynamical evolution of matter calculated. The existence of these locally inertial real coordinate systems and a local Y quantum field does not conflict with the Principle of Equivalence (except possibly in the strong extrapolated form: "In an arbitrary gravitational field, in the neighborhood of any space-time point it is possible to specify a locally inertial coordinate system such that *all* laws of Nature take the same form as in unaccelerated Cartesian coordinate systems in the absence of gravitation.") This strong formulation is not experimentally proven. We would rather require:

Two-Tier Strong Principle of Equivalence

In an arbitrary gravitational field, in the neighborhood of any space-time point, it is possible to specify a local inertial coordinate system such that all laws of *matter* take the same form as they would in unaccelerated Cartesian coordinate systems in the absence of gravitation, with the exception of the laws governing the imaginary quantum dimension(s) generated by the Y quantum field.

The fact that our unified theory of the known forces of Nature *self-consistently* has a weak gravitational field at high energies (the graviton sector is finite to all orders in perturbation theory) supports the formulation of eq. 9.12-3. Gravity becomes weaker at ultra-short distances. Therefore space-time is not quantum foam at ultra-short distances but rather smooth and flat a là special relativity – consistent with our formulation.

Quantum Gravity – Scalar Particle Model Lagrangian

While the application of the two-tier approach to the unified theory is a straightforward extension of the concepts and approaches described in the preceding chapters, it is useful to consider a simplified model that minimizes the tensorial verbiage so that the concepts and features might better stand out. The procedure differs only in detail for the cases of gauge fields.

The introduction of spinor fields would require the use of a two-tier vierbein formalism, which is straightforward to develop. The vierbein field $e^{\mu}{}_{a}$ would be a function of X, $e^{\mu}{}_{a}(X)$, with $g_{\mu\nu} = e_{\mu a}(X)e_{\nu}{}^{a}(X)$ where the index a is an index of a flat tangent space defined at each space-time point. The two-tier formulation of a vierbein theory is similar to the other two-tier formulations that we have considered and will not be developed here.

Thus we will consider the Lagrangian model for a scalar particle field interacting with the $g_{\mu\nu}$ gravitational field:

$$L_{GS} = \int d^4y \sqrt{g(X)}\, \mathscr{L}_{GS} + \mathscr{L}_C \qquad (9.16)$$

$$\mathscr{L}_{GS} = J\ \mathscr{L}_F{}^{Grav}(X^\mu) + J\ \mathscr{L}_{F\phi}(X^\mu) \tag{9.17}$$

with

$$\mathscr{L}_{F\phi} = ½\,[\,g^{\mu\nu}\partial\phi/\partial X^\mu\ \partial\phi/\partial X^\nu - m^2\phi^2] + \mathscr{L}_{F\phi\text{int}} \tag{9.18}$$

$$\mathscr{L}_{F\phi\text{int}} = \tfrac{1}{4!}\,\chi_0\ \phi(X(y))^4 + ½\,(m^2 - m_0{}^2)\phi^2 \tag{9.19}$$

which are covariant versions of eqs. 5.21 and 5.24.

A Justifiable Weak Field Approximation for Quantum Gravity

Many discussions of quantizing conventional gravity make a weak field approximation for the gravity sector which, in view of divergences in the resulting quantum field theory, are impossible to justify:

$$g_{\mu\nu} = \eta_{\mu\nu} + \kappa h_{\mu\nu} \tag{9.20}$$

where $\eta_{\mu\nu}$ is the flat space-time metric and $h_{\mu\nu}$ is a "small" deviation ($|h_{\mu\nu}| \ll 1$) from the flat space-time metric.

We shall see that the two-tier formulation of quantum gravity is finite and the effective field becomes increasingly weaker at short distances. Thus the weak field approximation becomes *more accurate* at short distances:

$$g_{\mu\nu}(X(y)) \simeq \eta_{\mu\nu} + \kappa h_{\mu\nu}(X(y)) \tag{9.21a}$$

At short distances space-time can be considered approximately flat (except possibly in the neighborhood of singularities) with quantum fluctuations embodied in $h_{\mu\nu}$. Thus eq. 9.21a is reasonable within the context of Two-Tier Quantum Gravity.

To first order in $h_{\mu\nu}$ the square root of the absolute value of the determinant of the metric tensor is:

$$\sqrt{g(X)} \simeq 1 + ½\,\kappa h^\sigma{}_\sigma(X(y)) \tag{9.21b}$$

Quantization of Quantum Gravity – Scalar Particle Model

We now proceed to quantize gravity in a straightforward way based on the linearization of the gravitational field equations in the weak field approximation. Assuming eq. 9.21a and keeping terms to first order in $h_{\mu\nu}$ gives the affine connection:

$$_X\Gamma^\sigma{}_{\mu\nu} = ½\,\kappa\eta^{\sigma\alpha}[\partial h_{\alpha\nu}/\partial X^\mu + \partial h_{\alpha\mu}/\partial X^\nu - \partial h_{\mu\nu}/\partial X^\alpha] + \mathcal{O}(h^2) \tag{9.22}$$

and the Ricci tensor:

$$ {}_XR_{\mu\nu} = \partial_X\Gamma^\lambda{}_{\lambda\mu}/\partial X^\nu - \partial_X\Gamma^\lambda{}_{\mu\nu}/\partial X^\lambda + \mathcal{O}(h^2) \qquad (9.23) $$

Thus the gravitation lagarangian terms become

$$ L^{Grav} = \int d^4y\, \sqrt{g(X)}\, J\mathscr{L}_F{}^{Grav}(X^\mu) \to L^{Grav}{}_{linear} = \int d^4y\, J\mathscr{L}^{Grav}{}_{linear}(X^\mu) \qquad (9.24) $$

The scalar particle Lagrangian terms become

$$ L^\phi = \int d^4y\sqrt{g(X)}J\mathscr{L}_{F\phi} \to L^\phi{}_{linear} = \int d^4yJ\{[½(\eta^{\mu\nu}\partial_\mu\phi\partial_\nu\phi - m^2\phi^2) + \mathscr{L}_{F\phi int}] + $$

$$ + ½\kappa h^{\mu\nu}\partial_\mu\phi\partial_\nu\phi + ¼\,\kappa h(\eta^{\mu\nu}\partial_\mu\phi\partial_\nu\phi - m^2\phi^2) + $$

$$ + ½\,\kappa h\mathscr{L}_{F\phi int}\} \qquad (9.25) $$

with the notation $h = h^\sigma{}_\sigma$ and using

$$ \partial_\mu \equiv \partial/\partial X^\mu \qquad (9.26) $$

$\eta^{\mu\nu}$ and $\eta_{\mu\nu}$ are used to raise and lower indices in keeping with the linearized, weak field approximation.

The Y terms in the Lagrangian can be written as:

$$ L^Y = \int d^4y\mathscr{L}_C = -¼\int d^4y\eta^{\mu\nu}\eta^{\alpha\beta}F_{Y\mu\alpha}F_{Y\nu\beta} \qquad (9.27) $$

We will lump the higher order terms (in h) in the gravity part of the Lagrangian, and the scalar particle part of the Lagrangian, into

$$ L_{Higher} = \int d^4y\, J\mathscr{L}_{Higher}(h, \phi) \qquad (9.28) $$

Thus the complete lagragian for a scalar particle interacting with gravitons is

$$ L_{GS} = L^{Grav}{}_{linear} + L^\phi{}_{linear} + L^Y + L_{Higher} \qquad (9.29) $$

The linearized gravitational Lagrangian term $L^{Grav}{}_{linear}$ generates the field equations:

$$\Box_X h_{\mu\nu} + \partial_\nu\partial_\mu h - \partial_\alpha\partial_\nu h^\alpha{}_\mu - \partial_\alpha\partial_\mu h^\alpha{}_\nu = \kappa S_{\mu\nu} \qquad (9.30)$$

where

$$\partial_\mu S^\mu{}_\nu = ½\, \partial_\nu S^\sigma{}_\sigma \qquad (9.31)$$

to 0th order in h and where

$$\Box_X = (\partial/\partial X^\nu)(\partial/\partial X_\nu) \qquad (9.32)$$

The most general coordinate transformation that maintains the weakness of the gravitational field has the form:

$$y^\alpha \rightarrow \; y'^\alpha = y^\alpha + \chi^\alpha(X(y)) \qquad (9.33)$$

This transformation induces a gauge transformation in $h_{\mu\nu}$ to:

$$h'_{\mu\nu} = h_{\mu\nu} - \partial_\mu \chi_\nu - \partial_\nu \chi_\mu \qquad (9.34)$$

It is easy to verify that eq. 9.30 is satisfied by $h'_{\mu\nu}$ if it is satisfied by $h_{\mu\nu}$.

Let us assume that we perform a gauge transformation making $h_{\mu\nu}$ traceless:

$$h^\sigma{}_\sigma = 0 \qquad (9.35)$$

and choose the gauge

$$\partial^\mu h_{\mu\nu} = 0 \qquad (9.36)$$

then eq. 9.30 becomes the wave equation:

$$\Box_X h_{\mu\nu} = \kappa S_{\mu\nu} \qquad (9.37)$$

Another gauge transformation of the free field $h_{\mu\nu}$ ($S_{\mu\nu} = 0$) makes

$$h_{\mu 0} = h_{0\mu} = 0 \qquad (9.38)$$

while retaining

$$h_{\mu\nu} = h_{\nu\mu} \tag{9.39}$$

The general solution[37] of the free field $h_{\mu\nu}$ (with $S_{\mu\nu}= 0$ in eq. 9.37) can be expressed as a fourier expansion:

$$h_{\mu\nu}(X(y)) = \int d^3k\, N_0(k) \sum_{\lambda=1}^{2} \varepsilon_{\mu\nu}(k, \lambda)[a(k,\lambda)\, e^{-ik.X} + a^\dagger(k,\lambda)\, e^{ik.X}] \tag{9.40}$$

where $\lambda = 1,2$ labels the ±2 helicity states, and where $N_0(k)$ is specified by eq. 3.25. The equal time ($y'^0 = y^0$) commutation relations are:

$$[h_{\mu\nu}(X(y)), h_{\alpha\beta}(X(y'))] = [\pi_{\mu\nu}(X(y)), \pi_{\alpha\beta}(X(y'))] = 0 \tag{9.41}$$

$$[h_{\alpha\beta}(X(y')), \pi_{\mu\nu}(X(y))] = i\, \mathscr{D}_{\alpha\beta,\mu\nu}(\mathbf{X}(y) - \mathbf{X}(y')) \tag{9.42}$$

for μ, $\nu = 1, 2, 3$ and where

$$\pi_{\mu\nu}(X(y)) = \partial h_{\mu\nu}(X(y))/\partial y^0 \tag{9.43}$$

in the Y Coulomb gauge where $X^0 = y^0$. $\mathscr{D}_{\alpha\beta,\mu\nu}$ is specified by:

$$\mathscr{D}_{\alpha\beta,\mu\nu}(X(y)- X(y')) = \int d^3k\, e^{i\,\mathbf{k}\bullet(X(y) - X(y'))}\, \Pi_{\alpha\beta\mu\nu}(\mathbf{k})/(2\pi)^3 \tag{9.44}$$

$$\Pi_{\alpha\beta\mu\nu}(\mathbf{k}) = ½\, [(\delta_{\alpha\mu} - k_\alpha k_\mu/\mathbf{k}^2)(\delta_{\beta\nu} - k_\beta k_\nu/\mathbf{k}^2) + (\delta_{\alpha\nu} - k_\alpha k_\nu/\mathbf{k}^2)(\delta_{\beta\mu} - k_\beta k_\mu/\mathbf{k}^2) -$$

$$- (\delta_{\alpha\beta} - k_\alpha k_\beta/\mathbf{k}^2)(\delta_{\mu\nu} - k_\mu k_\nu/\mathbf{k}^2)] \tag{9.45}$$

where α, β, μ, ν = 1, 2, 3.

The "transverse" graviton propagator can be represented as a time-ordered product of field operators:

$$i\Delta_{F2}{}^{TT}(y_1 - y_2)_{\lambda\tau\rho\sigma} = <0|T(h_{\lambda\tau}(X(y_1)), h_{\rho\sigma}(X(y_2)))|0> \tag{9.46}$$

[37] S. Weinberg, Phys. Rev. **135**, B1049 (1964); Phys. Rev. **138**, B988 (1965)

$$= -i \int \frac{d^4k\, e^{-ik\bullet(y_1 - y_2)}\, b_{\lambda\tau\rho\sigma}(k) R(\mathbf{k}, y_1 - y_2)}{(2\pi)^4\, (k^2 + i\varepsilon)}$$

where $R(\mathbf{k}, y_1 - y_2)$ is the gaussian factor appearing in propagators throughout two-tier theories and where $b_{\mu\nu\rho\sigma}(k)$ is a function of k only:

$$b_{\alpha\beta\mu\nu}(k) = ½[(\eta_{\alpha\mu} - k_\alpha k_\mu/\mathbf{k}^2)(\eta_{\beta\nu} - k_\beta k_\nu/\mathbf{k}^2) + (\eta_{\alpha\nu} - k_\alpha k_\nu/\mathbf{k}^2)(\eta_{\beta\mu} - k_\beta k_\mu/\mathbf{k}^2) -$$

$$- (\eta_{\alpha\beta} - k_\alpha k_\beta/\mathbf{k}^2)(\eta_{\mu\nu} - k_\mu k_\nu/\mathbf{k}^2)] \quad (9.47)$$

where $\alpha, \beta, \mu, \nu = 0, 1, 2, 3$.

The quantum gravitational interaction also has an "instantaneous" part (similar to the instantaneous Coulomb interaction of QED) in addition to the transverse interaction embodied in eq. 9.46. This "instantaneous" interaction contains the Newtonian potential (described later). The sum of the instantaneous interaction and the transverse interaction gives the total gravitational interaction.

The above graviton propagator has the form given in eq. 6.102. The caculation of the leading behavior is the same as that of the two-tier scalar boson propagator except for the presence of factors such as $\eta_{\rho\sigma}$. The leading momentum dependence of the graviton propagator in momentum space is

$$i\Delta_{F2}{}^{TT}(p)_{\lambda\tau\rho\sigma} \backsim p^{-6} \quad (9.48)$$

The graviton vertices in two-tier Quantum Gravity will be described within the framework of the path integral formulation.

Quantum Gravity–Scalar Particle Model Path Integral

A path integral formalism can be developed for two-tier Quantum Gravity interacting with matter fields. In this section we will consider the case of a matter field consisting of massive scalar bosons with a quartic interaction. The path integral formalism that we develop is similar to that of Yang-Mills theories in the previous chapter.

The two-tier path integral for a Quantum Gravity–Scalar Particle Theory can be written as:

$$Z(J, J^{\mu\nu}) = N \int D\phi Dh DY \Delta_{FPG}(h)\delta(F(h)) \exp\{i\int d^4y[\mathscr{J}(\mathscr{L}^{Grav}{}_{linear}(X^\mu) +$$

$$+ \mathscr{L}^{\phi}{}_{\text{linear}}(X^{\mu}) + \mathscr{L}_{\text{Higher}}(h, \phi) + \mathscr{L}_{C}(X, y) +$$

$$+ j_{\mu}(y)Y^{\mu}(y) + J(y)\phi(X) + J^{\mu\nu}(y)h_{\mu\nu}(X)]\}\Big|_{j_{\mu} = 0} \qquad (9.49)$$

where $\delta(F(h))$ specifies the gauge as a functional delta function, and $\Delta_{FPG}(h)$ is the corresponding Fadeev-Popov determinant. $\mathscr{J}$ is the Jacobian for the transformation from y coordinates to X coordinates. The Fadeev-Popov determinant $\Delta_{FPG}(h)$ can be calculated in the standard way. First we note

$$\delta(F(h^{\chi})) = \delta(\chi - \chi_0)\ \left|\det \delta F(h_{\mu\nu}{}^{\chi}(X))/\delta\chi(X)\right|^{-1}\Big|_{F(h)=0} \qquad (9.50)$$

where

$$h_{\mu\nu}{}^{\chi} = h_{\mu\nu} - \partial_{\mu}\chi_{\nu} - \partial_{\nu}\chi_{\mu} \qquad (9.34)$$

Then

$$\Delta_{FPG}(h) = \left|\det \delta F(h^{\chi}(X))/\delta\chi(X)\right|\Big|_{F(h)=0} \qquad (9.51)$$

We will choose the gauge of eq. 9.36 to evaluate the Fadeev-Popov determinant. Under an infinitesimal gauge transformation of the form:

$$h_{\mu\nu}{}^{\chi}(X) = h_{\mu\nu}(X) - \partial_{\mu}\chi_{\nu} - \partial_{\nu}\chi_{\mu} \qquad (9.52)$$

which preserves the weak field nature of $h_{\mu\nu}$, we find

$$F_{\nu}(h^{\chi}) = \partial^{\mu}(h_{\mu\nu}(X) - \partial_{\mu}\chi_{\nu} - \partial_{\nu}\chi_{\mu})$$

$$= -\Box_X\, \chi_{\nu}(X) - \partial_{\nu}\partial^{\mu}\chi_{\mu} \qquad (9.53)$$

Thus

$$\delta F_{\mu}(h^{\chi}(X))/\delta\chi^{\nu}(X) = -\eta_{\mu\nu}\Box_X - \partial_{\mu}\partial_{\nu} \qquad (9.54)$$

and

$$\Delta_{FP}(A) = \left|\det\left(-\eta_{\mu\nu}\Box_X - \partial_\mu\partial_\nu\right)\right|\Big|_{F(h)=0} \qquad (9.55)$$

We note the two-tier Fadeev-Popov determinant is solely a function of the X coordinates. The determinant only introduces an overall multiplicative constant that can be absorbed into the normalization constant N. This fact becomes evident if we follow the standard procedure and rewrite the determinant as a path integral over anti-commuting c-number fields with a ghost Lagrangian. Then we see that the ghost does not interact with the other fields and thus only generates an overall multiplicative constant that can be absorbed in N:

$$\Delta_{FPG}(h) = \int Dc^*Dc \exp[\, i\int d^4X\, \mathscr{L}^{ghost}(X^\mu)] \qquad (9.56)$$

where

$$\mathscr{L}^{ghost}(X^\mu) = c^{\mu*}(X)[\eta_{\mu\nu}\Box_X + \partial_\mu\partial_\nu]c^\nu(X) \qquad (9.57)$$

We now go through the same analysis as we did in the ϕ^4 theory path integral example and the Yang-Mills path integral example (with some superficial differences). First we integrate the linear part of the Y field Lagrangian as we did previously. Then we integrate the linear part of the ϕ field Lagrangian as done previously. Lastly we integrate the linear part of the gravitation Lagrangian to obtain the path integral for the perturbative expansion with the result:

$$Z(J, J^{\mu\nu}) = N\Big\{\exp\big[i\int d^4y \mathscr{L}_{Higher}(-i\delta/\delta J^{\mu\nu}(y), -i\delta/\delta J(y))\big]\cdot$$

$$\exp[-\tfrac{1}{2}\, i\int d^4y_1 d^4y_2\, J^{\mu\nu}(y_1)\Delta_{F2}{}^{TT}(y_1 - y_2, z)_{\mu\nu\rho\sigma}J^{\rho\sigma}(y_2)]\cdot$$

$$\exp[-\tfrac{1}{2}\, i\int d^4y_1 d^4y_2\, J(y_1)\Delta_F{}^{TT}(y_1 - y_2, z)J(y_2)]\Big\}\Big|_{z=y_1-y_2} \qquad (9.58)$$

There are two issues that arise in the development of eq. 9.58: 1.) The integral over y in $\int d^4y \mathscr{L}_{Higher}$ which began as the integral $\int d^4y \mathscr{J} \mathscr{L}_{Higher} = \int d^4X \mathscr{L}_{Higher}$ in eq. 9.49; and 2.) The handling of derivatives with respect to X in $\mathscr{L}_{Higher}$.

1.) See the discussions following eqs. 6.34 and 8.29 that apply here as well without change.

2.) See the discussion following eq. 8.32, which applies here without change. In particular we note that the derivative with respect to X of the graviton propagator (eq. 9.46-7) is specified by the following:

$$\partial i\Delta_{F2}{}^{TT}(y_1 - y_2)_{\lambda\tau\rho\sigma}/\partial X^{\mu}(y_1) = \partial[i\Delta_{F2}{}^{TT}(y_1 - y_2,\, z)_{\lambda\tau\rho\sigma}]/\partial y_1{}^{\mu}\,\Big|_{z = y_1 - y_2} \tag{9.59}$$

where

$$i\Delta_{F2}{}^{TT}(y_1 - y_2,\, z)_{\lambda\tau\rho\sigma} = -i \int \frac{d^4k\, e^{-ik\bullet(y_1 - y_2)}\, b_{\lambda\tau\rho\sigma}(k)R(\mathbf{k}, \mathbf{z})}{(2\pi)^4\,(k^2 + i\varepsilon)} \tag{9.60}$$

Thus

$$\frac{\partial\, i\Delta_{F2}{}^{TT}(y_1 - y_2)_{\lambda\tau\rho\sigma}}{\partial X^{\mu}(y_1)} = -i \int \frac{d^4k\, e^{-ik\bullet(y_1 - y_2)}\,(-ik_{\mu})b_{\lambda\tau\rho\sigma}(k)R(\mathbf{k}, y_1 - y_2)}{(2\pi)^4\,(k^2 + i\varepsilon)} \tag{9.61}$$

Therefore derivatives with respect to X in the interaction Lagrangian terms can be replaced by derivatives with respect to y if the graviton propagator is generalized to eq. 9.60. After taking all derivatives with respect to y, we set z equal to the respective $y_1 - y_2$ (actually the difference of the appropriate variables) in each propagator with results similar to eq. 9.61. The net consequence is that graviton vertices result in exactly the same polynomials in momenta as the conventional theory.

Thus two-tier gravity generates a perturbative expansion identical to conventional quantum gravity except that each graviton propagator has a gaussian damping factor $R(\mathbf{k},y_1 - y_2)$. At low energies the tree diagrams of conventional gravity theory emerge to good approximation in two-tier gravity. All diagrams with loops converge. Thus two-tier gravity is finite.

Finiteness of Quantum Gravity–Scalar Particle Model

Two-tier Quantum Gravity perturbation theory is finite. Calculations are highly convergent at large momentum ($\gtrsim M_c$). At low momentum the two-tier theory is similar to conventional gravity – particularly for tree diagrams and convergent diagrams.

For pure *conventional* Quantum Gravity DeWitt[38] finds the superficial degree of divergence of a diagram to be:

$$D = -2L_i + 2\sum_n V_n + 4K \qquad (9.62)$$

where L_i is the number of internal lines, V_n is the number of n-pronged vertices, and K is the number of independent momentum integrations. DeWitt further points out

$$K = L_i - \sum_n V_n + 1 \qquad (9.63)$$

Thus the superficial degree of divergence of a *conventional* Quantum Gravity diagram is:

$$D = 2(K + 1) \qquad (9.64)$$

for $K \geq 1$, displaying an ever increasing degree of divergence as the order of the diagram increases.

In the case of *Two-Tier Quantum Gravity* the superficial degree of divergence of a diagram is:

$$D_{TT} = -6L_i + 2\sum_n V_n + 4K \qquad (9.65)$$

(from eq. 9.48) with the result (taking account of eq. 9.63):

$$D_{TT} = -2L_i - 2\sum_n V_n + 2 \qquad (9.66)$$

Since any diagram with a loop has $L_i \geq 1$ and $\sum_n V_n \geq 1$ we see that $D \leq -2$. Thus *all* diagrams are convergent and *the two-tier formulation of Quantum Gravity theory is finite. The addition of arbitrary species of other two-tier fields – matter and gauge fields – does not introduce divergences in the combined two-tier theory.*

Unitarity of Quantum Gravity–Scalar Particle Model

The two-tier Quantum Gravity – Scalar Particle Model *superficially* appears to have a unitarity problem due to the non-hermitean nature of its hamiltonian. The lack of hermiticity is due entirely to the appearance of iY^μ in the X^μ field coordinates.

---

[38] B. S. DeWitt, Phys. Rev. **162**, 1239 (1967).

Thus interaction Lagrangian is not hermitean:

$$L_{Higher} = \int d^3y' \mathscr{L}_{Higher}(y' + iY(y')/M_c^2) \tag{9.67}$$

but

$$L_{Higher}^{\dagger} = \int d^3y' \mathscr{L}_{Higher}(y' - iY(y')/M_c^2) \neq L_{Higher} \tag{9.68}$$

The relation between L_{Higher} and its hermitean conjugate is

$$L_{Higher} = V\, L_{Higher}^{\dagger}\, V \tag{9.69}$$

where $V^2 = I$ is the metric operator defined in eqs. 5.16 – 5.18. By eq. 6.37 we see as a result that the two-tier S matrix is not unitary – it is pseudo-unitary:

$$S^{-1} = V\, S^{\dagger}\, V \tag{9.70}$$

Therefore

$$S^{\dagger} V S = V \tag{9.71}$$

We will now show that the S matrix is effectively unitary between physical asymptotic states – states consisting of only scalar ϕ particles and gravitons.

As we noted in an earlier discussion of coherent states, the S matrix is *block diagonal* in the sense that in-states consisting of only physical particles will only produce out-states consisting of physical particles. The consequence of block diagonality is that *S is unitary within the physical states sector*. To prove this point consider an S matrix element between physical asymptotic states:

$$<f\ in|V|i\ in> = <f\ in|S^{\dagger}\ VS|i\ in> = \sum_{n,\, m} <f\ in|S^{\dagger}\ |n><n|V|m><m|S|i\ in> \tag{9.72}$$

where the block diagonality of S (and the diagonality of V) limits the intermediate states |n> and |m> to states containing physical particles only. We note V is diagonal with eigenvalue one between states consisting of physical particles only. Thus eq. 9.72 becomes

$$\delta_{fi} = \sum_n \langle f\ in|S^\dagger\ |n\rangle\langle n|S|i\ in\rangle \tag{9.73}$$

or

$$S_{phys}^\dagger S_{phys} = I$$

So that

$$S_{phys}^\dagger = S_{phys}^{-1} \tag{9.74}$$

proving unitarity for physical states.

The Mass Scale M_c

The mass scale of two-tier theories is set by M_c. This mass scale cannot be ascertained with any degree of certainty at current, experimentally accessible, accelerator energies. Cosmic ray data also does not seem to give any clues as to the value of M_c. It appears that M_c is probably above 10^3 GeV/c^2 and may be of the order of (or equal to) the Planck mass:

$$\sqrt{\hbar c/G} = 1.221\times10^{19}\ \text{GeV/c}^2 \tag{9.75}$$

If M_c is of the 1,000 GeV/c^2 or larger the differences between its predictions at current accelerator energies and the predictions of conventional renormalized perturbation theory will be negligible. Actually a much lower value of M_c would still be consistent with the current stringent QED theoretical predictions as well as other predictions of conventional renormalized perturbation theory.

Planck Scale Physics

A finite theory of Quantum Gravity can provide information on the issues that have been of concern for many years – including the short distance behavior of the gravitational metric and ultra-small black holes.

Quantum Foam

Some theorists have conjectured that the classical view of smooth, almost flat space-time does not hold in the quantum regime at energies of the order of the Planck mass. Suggestions that space-time dissolves into quantum foam have appeared.

The finite two-tier formulation of Quantum Gravity is well-behaved at short distances and suggests that the quantum behavior of gravity and space-time in the short distance limit does not have limitless quantum fluctuations that result in a foam-like space-time picture.

Measurement of the Quantum Gravity Field

A number of conceptual problems have been raised about the effects of quantized General Relativity. Two-tier Quantum Gravity seems to resolve these issues.

Measurement of Time Intervals

Wigner[39] has studied the measurement of time intervals in General Relativity and sees a problem in the measurement of extremely short intervals. According to Wigner: the measurement of a time inteval in a region of space requires the measurement of the length of time required for an event to happen. The measurement requires an accurate clock. But the accuracy of the clock is limited by the energy-time uncertainty relation:

$$\Delta E \Delta t \geq \hbar \tag{9.76}$$

Thus the uncertainty in the clock's time measurement is related to the uncertainty in the clock's energy which is, in turn, related to the uncertainty in the clock's mass:

$$\Delta E = (\Delta m)c^2 \tag{9.77}$$

To obtain "infinite" accuracy the uncertainty (fluctuations) in the clock's mass must be infinite and thus the clock's mass must be infinite. Infinite fluctuations in the clock's mass will produce corresponding infinite fluctuations in the gravitational field.

$$\Delta h \propto \Delta E \qquad \text{(in conventional General Relativity)} \tag{9.78}$$

As a result the notion of space-time and time intervals (which depend on the geometry through General Relativity) become uncertain. Thus, according to Wigner and others, the concept of time intervals and space-time points becomes questionable.

The two-tier version of Quantum Gravity offers a potential way out of this dilemma. The gravitational force becomes stronger as one goes to shorter distances (higher energies) down to a distance (up to an energy) whose scale is set by M_c. At shorter distances (higher energies) the gravitational force becomes weaker and declines to zero at zero distance. Thus at very high energy the gravitational field fluctuations (Δh) are at worst inversely proportional to the energy (and probably decline by a higher power of inverse energy.) (The same considerations would apply if one chooses to consider fluctuations in the Riemann-Christoffel symbols.)

$$\Delta h < c_1/E < c_1/(\Delta E) \qquad \text{(in two-tier Quantum Gravity)} \tag{9.79}$$

---

[39] E. P. Wigner, Rev. Mod. Phys. **29**, 255 (1957); J. Math. Phys. **2**, 207 (1961).

Thus Wigner's conclusion does not hold in the two-tier version of Quantum Gravity as gravitational fluctuations actually become smaller at energies above a critical energy whose scale is set by M_c.

In fact, combining eqs. 9.79 and 9.76 we see

$$c_1 \Delta t / \Delta h \geq \hbar \tag{9.80}$$

at sufficiently high energy. Therefore the time uncertainty Δt, and the gravitational field fluctuations Δh, can both decrease while maintaining the energy-time uncertainty relation. Thus the notion of a space-time point "is saved" in two-tier quantum gravity.

Vacuum Fluctuations in the Gravitation Fields

While the expectation value of the free graviton field $h_{\mu\nu\text{conv}}(X)$ is zero in a conventional quantum field theoric approach:

$$<0|h_{\mu\nu\text{conv}}(X)|0> = 0 \tag{9.81}$$

the vacuum fluctuations of the *conventional* quantum graviton field is quadratically divergent since

$$<0|h_{\mu\nu\text{conv}}(X)h_{\alpha\beta\text{conv}}(X)|0> = \int d^3p\, b'_{\mu\nu\alpha\beta}(p)/[(2\pi)^3\, 2\omega_p] = \infty \tag{9.82}$$

where $b'_{\mu\nu\alpha\beta}(p)$ is a rational function of the momentum p.

In "two-tier" quantum field theory we find

$$<0|h_{\mu\nu}(X)h_{\alpha\beta}(X)|0> = \int d^3p\, b'_{\mu\nu\alpha\beta}(p)\, e^{-p^ip^j\Delta_{Tij}(0)}/[(2\pi)^3 2\omega_p] = 0 \tag{9.83}$$

since the exponential factor in the integrand is $-\infty$. The exponent contains

$$\Delta_{Tij}(z) = \int d^3k\, e^{-ik\bullet z}(\delta_{ij} - k_ik_j/\mathbf{k}^2)/[(2\pi)^3 2\omega_k] \tag{4.8}$$

Thus the vacuum fluctuations of $h_{\mu\nu}$ are zero in "two-tier" quantum field theory.

The Two-Tier Gravitational Potential vs. Newton's Gravitational Potential

The familiar gravitational potential of Newton is:

$$V_{Newton} = -\,G/|\mathbf{r}| \tag{9.84}$$

The Two-Tier gravitational potential is:

$$V_{Two\text{-}Tier} = -\,G\Phi(M_c^2\pi|\mathbf{r}|^2)/|\mathbf{r}| \tag{9.85}$$

where $\Phi(y)$ is the error function.[40] It can be calculated in Two-Tier Quantum Gravity from Two-Tier Quantum Gravity propagator terms similar to corresponding terms in the Two-Tier photon propagator that led to the Two-Tier Coulomb potential (eqs. 7.48 – 7.51). At small distances ($\pi r^2 \ll M_c^{-2}$)

$$V_{Two\text{-}Tier} \rightarrow -\,G2\sqrt{\pi}\, M_c^2|\mathbf{r}| \tag{9.86}$$

a linear potential, and at large distances ($\pi r^2 \gg M_c^{-2}$)

$$V_{Two\text{-}Tier} \rightarrow V_{Newton} = -\,G/|\mathbf{r}| \tag{9.87}$$

the Newtonian potential.

The Two-Tier gravitational potential has a minimum at

$$M_c^2\pi r_{MIN}^2 = 1 \tag{9.88}$$

At the minimum $V_{Two\text{-}Tier}$ has the value:

$$V_{Two\text{-}TierMIN} = -\,.8427G\sqrt{\pi}\, M_c \tag{9.89}$$

Figs.9.1 – 9.3 display plots of $V_{Two\text{-}Tier}$ for $M_c = 10^5$ GeV/c^2, $M_c = 10^{12}$ GeV/c^2 , and $M_c = 1.22\ 10^{19}$ GeV/c$^2 = G^{-\frac{1}{2}}$ – the Planck mass. Distances and energies are stated in terms of Gev^{-1} and GeV respectively. They may be converted to cgs units in the standard way.

[40] W. Magnus and F. Oberhettinger, *Formulas and Theorems for the Special Functions of Mathematical Physics* (Chelsea Publishing Co., New York, 1949) page 96.

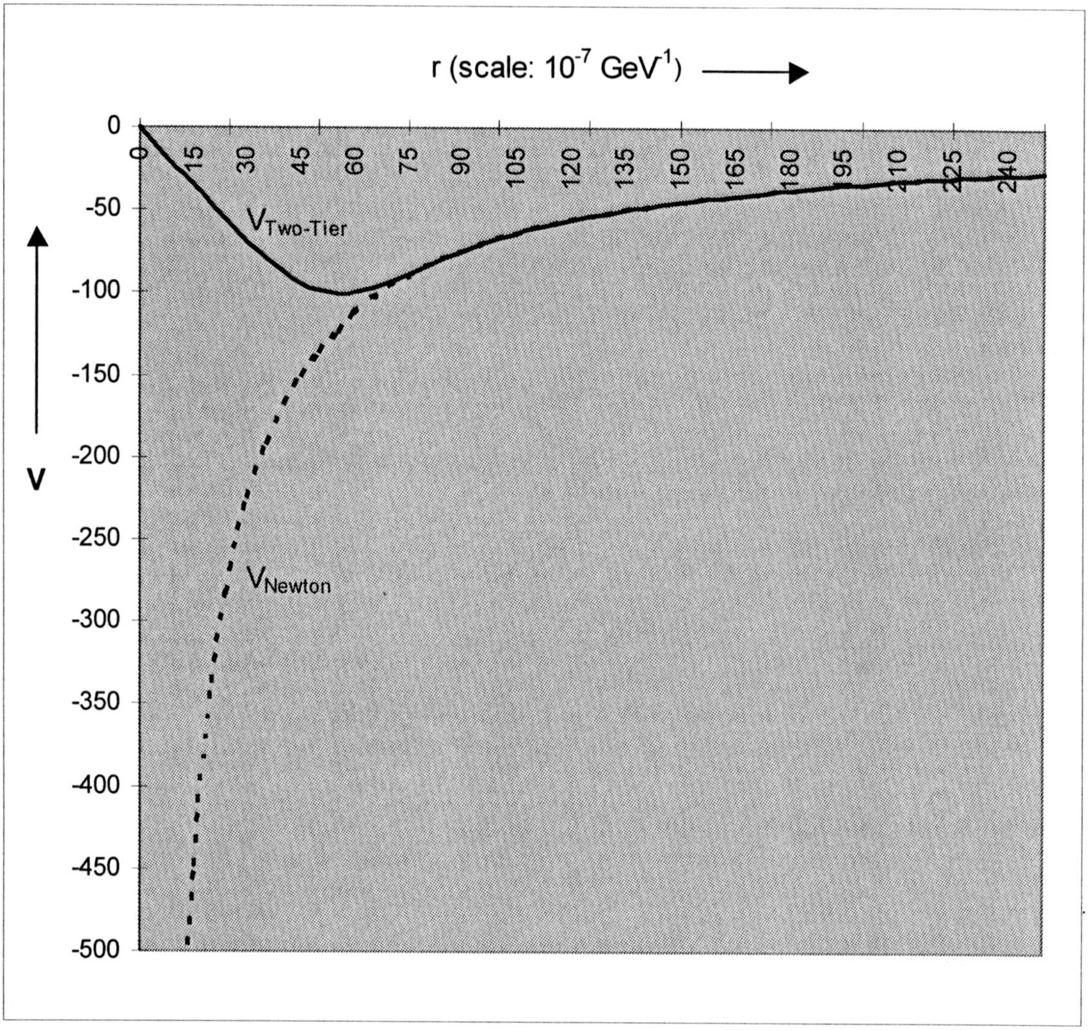

Figure 9.1. Plot of Two-Tier gravitational potential for $M_c = 10^5$ GeV/c^2 and Newton's gravitational potential. The potentials are measured in units of 10^{-33} GeV. The radial distance is measured in units of 10^{-7} GeV^{-1}.

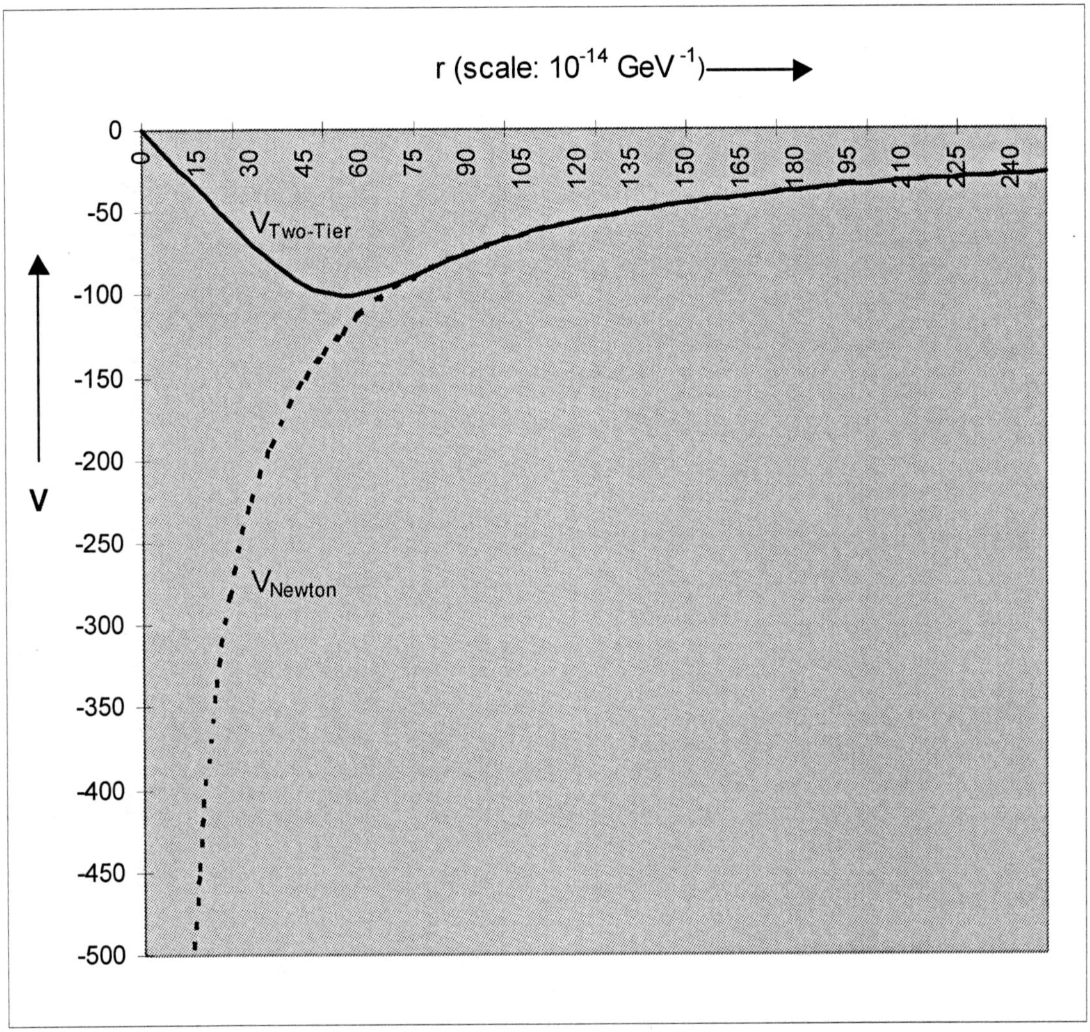

Figure 9.2. Plot of Two-Tier gravitational potential for $M_c = 10^{12}$ GeV/c^2 and Newton's gravitational potential. The potentials are measured in units of 10^{-26} GeV. The radial distance is measured in units of 10^{-14} GeV^{-1}.

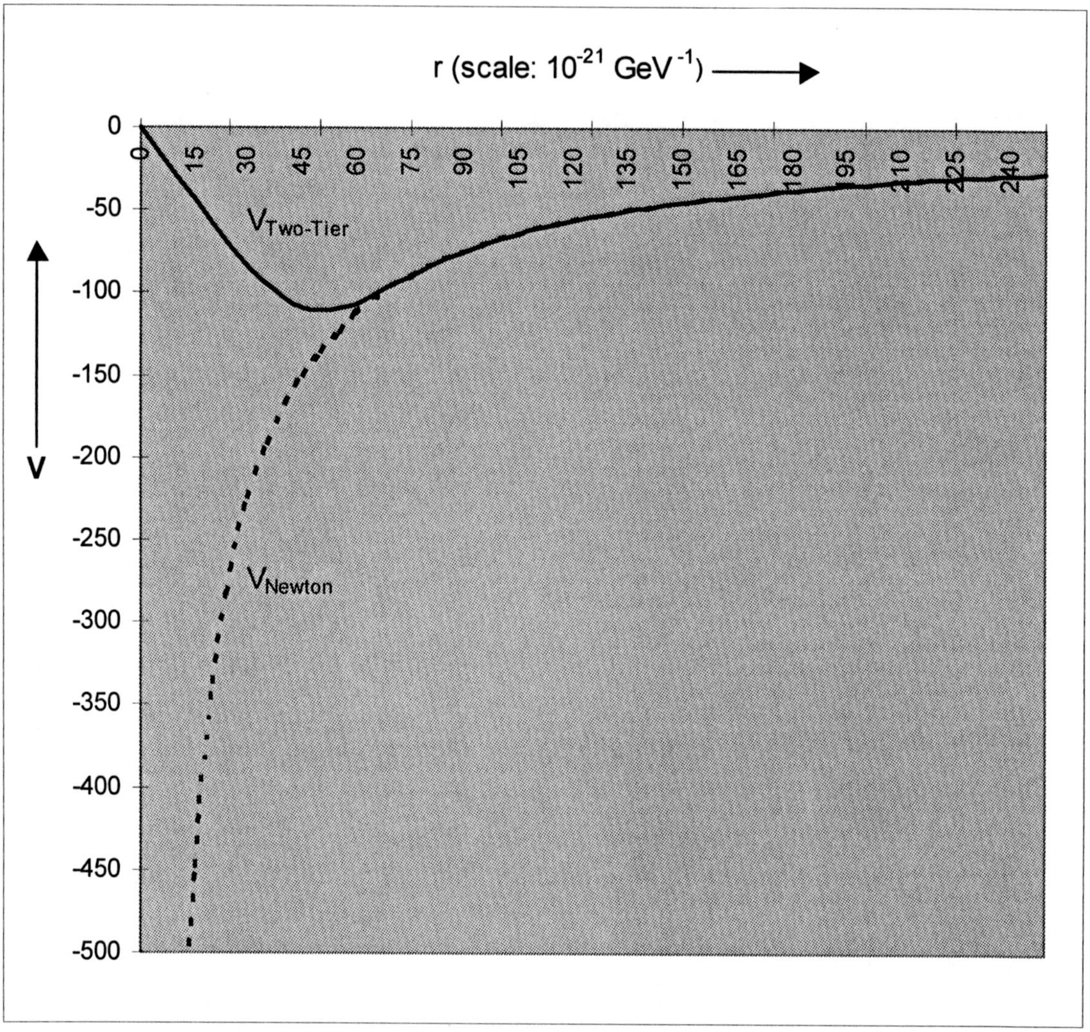

Figure 9.3. Plot of Two-Tier gravitational potential for M_c = 1.22 × 10^{19} GeV/c^2 (the Planck mass) and Newton's gravitational potential. The potentials are measured in units of 10^{-19} GeV. The radial distance is measured in units of 10^{-21} GeV^{-1}.

Black Holes

The existence of microscopic black holes has been the subject of much speculation. It appears that arbitrarily small black holes can exist in classical General Relativity. The divergences associated with the short distance behavior of its conventional quantization raise the possibility of additional singular behavior at short distances as well.

On the other hand, in two-tier Quantum Gravity, at short distances, when the distance scale becomes less than M_c^{-1} (and thus the energy scale becomes greater than M_c), the two-tier gravitational force grows smaller and become zero in the limit of zero distance or infinite energy. The preceding figures (Figs. 9.1 – 9.3) show the two-tier gravitational potential linearly approaches zero at short distances unlike the Newtonian gravitational potential which approaches $-\infty$ as r approches zero. Thus the short distance behavior of two-tier gravity suggests that black holes of ultra-small size may not exist in two-tier Quanum Gravity.

If we examine the two-tier gravitational potential we note that it is similar to the Newtonian potential until the separation distance approaches the minimum of the potential. Thus we might expect that conventional classical General Relativity would be approximately valid down to distances of the order of the location of the minimum of the two-tier potential. Based on this assumption and on the assumption that M_c equals the Planck mass:

Assumption: $$M_c = M_{Planck} = G^{-1/2} \tag{9.90}$$

we can calculate the mass of a black hole whose radius equals the minimum of the two-tier potential. From eq. 9.88 we obtain

$$r_{MIN} = (G/\pi)^{1/2} = r_{BlackHole} = 2GM_{BlackHoleMIN} \tag{9.91}$$

with the result

$$M_{BlackHoleMIN} = (4\pi G)^{-1/2} = \kappa^{-1} \tag{9.92}$$

by eq. 9.11 and

$$M_{BlackHoleMIN} \cong .282\ M_{Planck} \tag{9.93}$$

Since corrections to conventional quantum gravity are at most of the order of M_c^{-2} it appears that the value of $M_{BlackHoleMIN}$ is consistent with the approximate validity of classical expression for a black hole radius. We note

$$(M_{BlackHoleMIN}/M_c)^2 \cong .0795 \tag{9.94}$$

The decline of the two-tier gravitational potential at distances below M_c^{-1} suggests that there is a minimum size for black holes in Two-Tier Quantum Gravity.

10. A Unified Quantum Field Theory of the Known Forces of Nature

Formulation

The unification of QED and weak interactions in Electroweak Theory interrelated the theories within the framework of an overall SU(2)⊗U(1) symmetry and thus was significantly more then merely "glueing" the theories together.

The unification of Electroweak theory with Quantum Chromodynamics (QCD) into the Standard Model was a direct combination of these theories in which the symmetries of each respective theory were directly combined: SU(2)⊗U(1) symmetry from Electroweak Theory and SU(3) from QCD to produce the Standard Model with SU(2)⊗U(1)⊗SU(3) symmetry. Electroweak Theory was "glued together" with QCD to produce the Standard Model without an underlying rationale. Nevertheless, the Standard Model is a renormalizable theory that accounts for the vast majority of experimental data.

The present work has two goals: 1.) To make QED, Electroweak Theory, QCD and Quantum Gravity finite and thus remove a major long term defect in Quantum Field Theory, and 2.) To create a unified theory of the known forces of Nature from these pieces. Item 1 has been achieved in the preceding chapters using Two-Tier Quantum Field Theory. This chapter discusses item 2.

In this chapter we propose a finite unified theory of the Standard Model and Quantum Gravity within the framework of Two-Tier Quantum Field Theory. As in the case of the Standard Model our unified theory amounts to glueing together the Two-Tier version of the Standard Model with the Two-Tier version of Quantum Gravity. Therefore we regard the theory as provisional in the sense that a deeper unification remains to be formulated.

Nevertheless the Two-Tier Unified Theory may be of some importance beyond the satisfaction of having a finite theory of Nature. It might be a starting point for a deeper, more unified theory. It can possibly be used to address cosmological questions such as the state of the universe immediately after the Big Bang when the size of the universe was of the order of the Planck length or smaller. The interactions in the unified theory become weaker at short distances and thus low order perturbation theory becomes a better approximation to the exact results. The small size of the early universe will require the theory be reformulated with an appropriate c-number background metric. De Witt has developed such a formalism for conventional gravity. A similar formulation within the framework of Two-Tier Quantum Gravity is possible.

Unified Two-Tier Standard Model and Quantum Gravity Lagrangian

We define the Lagrangian, and action, for the two-tier unified quantum field theory of gravitation and the Standard Model as

$$L_{\text{Unified}} = \int d^4y\, \mathscr{L}_{\text{Unified}} \tag{10.1}$$

$$\mathscr{L}_{\text{Unified}} = J\sqrt{g(X)}\,\left(\mathscr{L}_F{}^{\text{Grav}}(X) + \mathscr{L}_F{}^{\text{SM}}(X)\right) + \mathscr{L}_C \tag{10.2}$$

with

$$\mathscr{L}_F{}^{\text{Grav}}(X) = (2\kappa^2)^{-1}{}_X R(X) \tag{10.3}$$

where $\mathscr{L}_F{}^{\text{SM}}$ is the complete "normal" Quantum Field theory Lagrangian for the Standard Model variant under consideration written in a general covariant form, g(X) is the absolute value of the determinant of $g_{\mu\nu}$, and J is the Jacobian of eq. 2.21. *All particle fields in* $\mathscr{L}_F{}^{\text{SM}}$ *are assumed to be functions of the X coordinate only. The dependence of the particle fields on the "underlying" coordinates* y^μ *is assumed to be solely through* X^μ. The Lagrangian $\mathscr{L}_{\text{Unified}}$ is a separable Lagrangian of the type of eq. 2.26 embodying the composition of extrema described in chapter 2.

As in all cases considered we define the coordinate part of the Lagrangian $\mathscr{L}_C$ as

$$\mathscr{L}_C = +\tfrac{1}{4}\, M_c{}^4 F^{\mu\nu}F_{\mu\nu} \tag{10.4}$$

with

$$F_{\mu\nu} = \partial X_\mu/\partial y^\nu - \partial X_\nu/\partial y^\mu \tag{10.5}$$

$$X_\mu(y) = y_\mu + i\, Y_\mu(y)/M_c{}^2 \tag{10.6}$$

Then we reexpress the Lagrangian as

$$\mathscr{L}_C = -\tfrac{1}{4}\, F_Y{}^{\mu\nu}F_{Y\mu\nu} \tag{10.7}$$

with

$$F_{Y_{\mu\nu}} = \partial Y_\mu / \partial y^\nu - \partial Y_\nu / \partial y^\mu \tag{10.8}$$

and

$$F_Y^{\mu\nu} = \eta^{\mu\alpha}\eta^{\nu\beta} F_{Y_{\alpha\beta}} \tag{10.9}$$

The development of the physics embodied in the Lagrangian proceeds along the lines described in the previous chapters and so will not be repeated here.

"Low Energy" Behavior

The low energy sector of the unified theory is defined as the sector with momenta whose values are much less than M_c. It is clear from the discussions of the previous chapters that the low energy sector of the unified theory is effectively identical to that of the corresponding conventional quantum field theory.

In addition the low energy behavior of the two-tier unified theory in the QED sector can closely approximate the results of QED calculations which have been found to agree well with experiment to an extremely high degree of accuracy.

Thus the two-tier unified theory satisfies a Correspondence Principle in the Standard Model sector: The low energy behavior of the two-tier Standard Model sector is the same as the behavior of the conventional Standard Model to a high degree of approximation. In addition the two-tier Quantum Gravity sector tree diagrams are the same as the conventional Quantum Gravity tree diagrams to a high degree of accuracy.

Negative Degree of Divergence – A Finite Unified Theory

The discussions of the perturbation theory of matter fields, gauge fields and gravitons show that the theory is finite.

Unitarity

The unitarity discussions of the various sectors of the unified theory in previous chapters show the unified theory to be unitary. As long as Y excitations are not allowed in in-states they will not appear in out-states. The S matrix is block diagonal and unitary within the physical asymptotic states sector.

Other General Issues

There are several other issues that might be a source of concern in the unified theory.

Does the Y Part of Lagrangian Breaks General Covariance?

It is clear from eq. 10.2 that $\mathscr{L}_C$ breaks the invariance of the action under general coordinate transformations. We have discussed this issue in detail in the section following eq.

9.15. To summarize, there is no justification for requiring the Y field sector to be general covariant.

Choice of Y Gauge Introduces Superficial Breaking of Special Relativity

We have chosen to work in the Coulomb gauge of the Y field throughout our discussions. The result is that unitarity is easier to verify and fictitious Y field quanta need not be introduced. One can develop a fully covariant special relativistic formulation using the Gupta-Bleuler gauge for the Y field. The results would be the same. Thus, although the formalism is not manifestly (special) relativistic, the results are consistent with Special Relativity.

REFERENCES

Akhiezer, N. I., Frink, A. H. (tr), *The Calculus of Variations* (Blaisdell Publishing, New York, 1962).

Bailin, D. & Love, A., *Supersymmetric Gauge Field Theory and String Theory* (Institute of Physics Publishing, Philadelphia, PA, 1994).

Bjorken, J. D., Drell, S. D., *Relativistic Quantum Fields* (McGraw-Hill, New York, 1965).

Bogoliubov N. N., & Shirkov, D. V., Volkoff, G. M. (tr), *Introduction to the Theory of Quantized Fields* (Wiley-Interscience, New York, 1959).

Gelfand, I. M., Fomin, S. V., Silverman, R. A. (tr), *Calculus of Variations* (Dover Publications, Mineola, NY, 2000).

Giaquinta, M., Modica, G., Souchek, J., *Cartesian Coordinates in the Calculus of Variations* Volumes I and II (Springer-Verlag, New York, 1998).

Giaquinta, M., Hildebrandt, S., *Calculus of Variations* Volumes I and II (Springer-Verlag, New York, 1996).

Goldstein H., *Classical Mechanics* (Addison-Wesley, Reading, MA 1965).

Huang, K., *Quarks, Leptons & Gauge Fields Second Edition* (World Scientific, River Edge, NJ, 1992).

Jost, J., Li-Jost, X., *Calculus of Variations* (Cambridge University Press, New York, 1998).

Kaku, M., *Introduction to Superstrings and M-Theory Second Edition* (Springer-Verlag, New York, 1999).

Kaku, M., *Quantum Field Theory* (Oxford University Press, New York, 1993).

Polchinski, J., *String Theory* (Cambridge University Press, New York, 1998).

Sagan, H., *Introduction to the Calculus of Variations* (Dover Publications, Mineola, NY, 1993).

Weinberg, S., *The Quantum Theory of Fields* (Cambridge University Press, New York, 1995).

INDEX

About the Author

Stephen Blaha is an internationally known physicist with extensive interests in Science, the Arts, and Technology. He has written a highly regarded book on physics, consciousness and philosophy – Cosmos and Consciousness, a book on Science and Religion entitled The Reluctant Prophets, a book applying physics to the history of civilizations, and books on Java and C++ programming. He has also developed a mathematical theory of civilizations that is described in his book, The Life Cycle of Civilizations. He has served on the faculties of Cornell University, Yale University, Syracuse University and Williams College. He was also a Member of the Technical Staff at Bell Laboratories, a Director of Wang Laboratories, and President of Blaha Software Inc. Dr. Blaha is noted for contributions to elementary particle theory and solid state physics theory as well as contributions to Computer Science. Among other achievements he was a co-discoverer of the "r potential" for heavy quark binding, first suggested the existence of topological structures in superfluid He-3, developed the first grammar-based formalism for quantum computers and applied it to elementary particle theories, and was a pioneer in UNIX development, database benchmarking, and Desktop Publishing.

Printed in the United States
1073400002B

9 780972 079594